EXERCISES IN
WORKSHOP MATHEMATICS
FOR YOUNG ENGINEERS

BY

LESLIE SMITH
B.Sc.

*Lecturer in Mathematics at the Derby and District College of Technology, and
Organizing Lecturer at the Works Training Schools,
British Railways, Derby*

W0042241

CAMBRIDGE
AT THE UNIVERSITY PRESS
1964

CAMBRIDGE UNIVERSITY PRESS
Cambridge, New York, Melbourne, Madrid, Cape Town,
Singapore, São Paulo, Delhi, Tokyo, Mexico City

Cambridge University Press
The Edinburgh Building, Cambridge CB2 8RU, UK

Published in the United States of America by Cambridge University Press, New York

www.cambridge.org
Information on this title: www.cambridge.org/9780521202954

© Cambridge University Press 1954

First edition 1954
Reprinted 1960, 1964
First paperback edition 2011

A catalogue record for this publication is available from the British Library

ISBN 978-0-521-06498-9 Hardback
ISBN 978-0-521-20295-4 Paperback

CONTENTS

PREFACE

The purpose of this book is to present a set of exercises of a practical nature to the young engineer, who is beginning to apply mathematical principles to the problems which confront him in the workshop.

The work is primarily designed for use in the Technical Secondary School, the County College and the Works Training School.

The author wishes to express his grateful thanks to Mr A. J. L. Avery, M.A., under whose inspiration this book has been produced, for constant help and advice; to Mr D. I. Robson, B.Sc., for the execution of the diagrams; and to the Railway Executive for the use of data appearing in the first exercise.　　　　　　　　　　　　　　　　　L. S.

ALLESTREE
DERBY
April 1954

FOREWORD

By E. J. LARKIN, A.M.I.Mech.E., M.I.Loco.E.

Director of Work Study,
British Transport Commission, London

Mr Smith is to be congratulated on publicizing such a representative selection of his vast collection of problems. Such a book could only be compiled by one who has a wealth of detailed knowledge of what the embryo craftsman requires to ensure systematic training.

Mr Smith has spent the whole of his life in the teaching profession and has nearly thirty years' experience to his credit. I believe he holds a unique position in being the first lecturer under a local education authority to be seconded to an industrial establishment to act as co-ordinating lecturer in workshop mathematics and associated subjects on behalf of the local education authority.

It is this combination of experience which has enabled the author to present fundamental arithmetical principles in the form of problems which arise in the workshop. The contents of the book are essentially practical.

As an individual, Mr Smith is recognized as being most successful in his profession, and all trainees, apprentices and others concerned with engineering training should find this book of tremendous interest and value.

Exercise 1. Locomotives: Lengths and Weights

1. Type: Power Class 8 P. Sir William A. Stanier, F.R.S. 4–6–2 No. 6256.

From a photograph by courtesy of British Railways

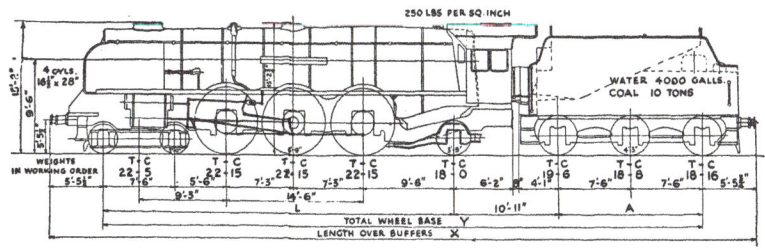

Fig. 1. 4–6–2 Passenger Engine (roller bearings)

Boiler: Barrel 20 ft. $3\frac{1}{16}$ in. Diam. outside 5 ft. $8\frac{5}{8}$ in. increasing to 6 ft. $5\frac{1}{2}$ in.
Firebox: Outside firebox 8 ft. 6 in. × 7 ft. $6\frac{5}{8}$ in. to 7 ft. $0\frac{1}{4}$ in.
Tubes: Superheater elements 5 P 4 type; large tubes 40 ft. $5\frac{1}{4}$ in. diam. outside × 7 s.w.g., small tubes 129 ft. $2\frac{3}{8}$ in. diam. outside × 11 s.w.g., both 19 ft. 3 in. between tube plates;
Heating surface: tubes 2577 sq.ft., firebox 230 sq.ft., superheater 979 sq.ft., total 2807 sq.ft.
Grate area: 50 sq.ft.
Tractive effort at 85 % *b.p.*: 40,000 lb.
Adhesion factor: 3·82.

From the data given in Fig. 1 calculate

(*a*) the length over buffers X;

(*b*) the distance L between the fore and rear bogey wheels;

(*c*) the total wheel base Y;

(*d*) the distance *A* between the front and rear wheel centres of the tender;

(*e*) the weight of the engine;

(*f*) the weight of the tender;

(*g*) the total weight of engine and tender;

(*h*) the ratio between the weight of the tender and the total weight.

From a photograph by courtesy of British Railways

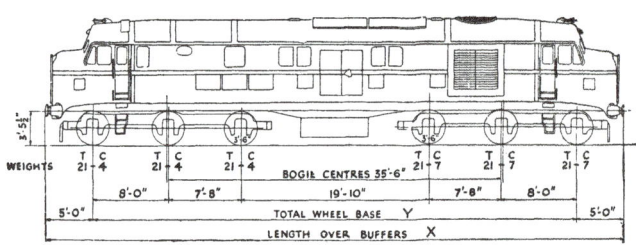

Fig. 2.   Main-Line Diesel Electric Locomotive

Engine: English Electric Type 16 S.V.T.; 16 cylinder 4-stroke Vee-type Diesel Engine with turbo-superchargers; 1600 h.p. at 750 r.p.m. 10 in. bore × 12 in. stroke.

Generator: English Electric Type E.E. 823A.

Traction Motors: (6) English Electric Type E.E. 519/1B.

Gear Ratio: 55/18.

Tractive effort: 41,400 lb.

Adhesion factor: 6·91.

Tank capacities: Engine fuel main, 815 gal.; Engine fuel service, 85 gal.; Radiator header, 40 gal.; Heating boiler feed water, 850 gal.; Heating boiler fuel oil, from main fuel.

2. Main Line Diesel Electric Locomotive.

From the data given in Fig. 2 calculate

(*a*) the length over buffers X;
(*b*) the total wheel base Y;
(*c*) the weight in working order;
(*d*) the number of gallons of fuel carried.

3. Calculate the clearance between the wheels of an Austerity Shunter 0–6–0, the wheels having a diameter of 4 ft. 3 in., if the distance between their centres is 5 ft. 9 in. and 5 ft. 3 in. respectively.

Exercise 2. Fractions: Addition and Subtraction

1. Calculate the length L of the gauge (Fig. 3).

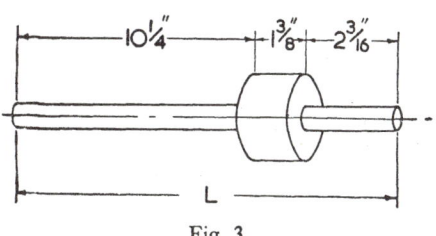

Fig. 3

2. Determine the length X of the worm shaft (Fig. 4).

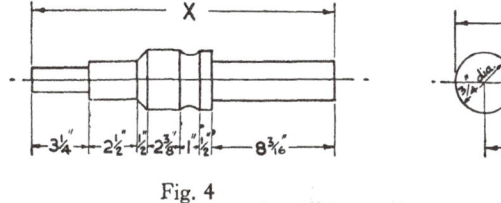

Fig. 4   Fig. 5

3. Determine the dimension d (Fig. 5).

4. Determine the dimension D (Fig. 6).

5. Calculate the length L of the lathe centre (Fig. 7).

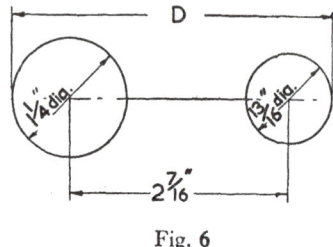

Fig. 6

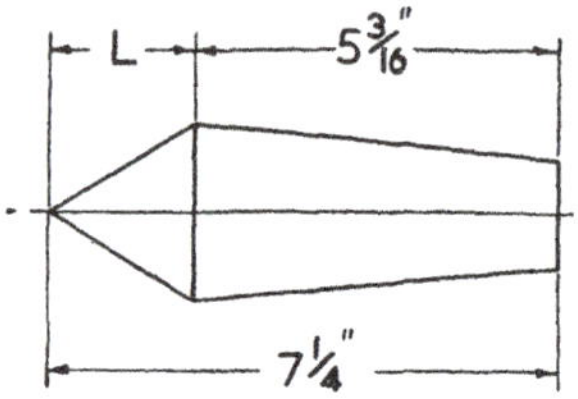

Fig. 7

6. Calculate the radius R of the spindle weight (Fig. 8).

7. Calculate the length L of the portion of the pin (Fig. 9).

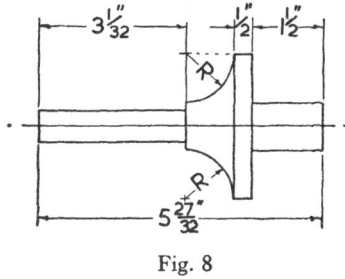

Fig. 8

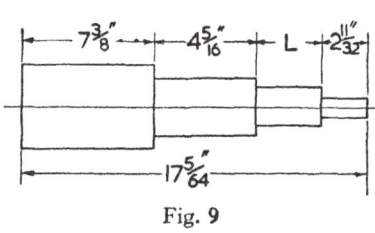

Fig. 9

8. Determine the dimension X (Fig. 10).

9. Calculate the length L (Fig. 11).

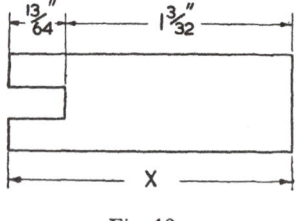

Fig. 10

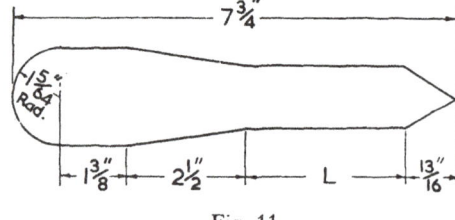

Fig. 11

10. Determine the diameter d (Fig. 12).

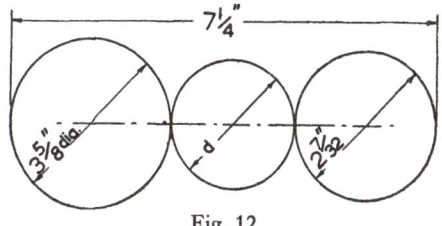

Fig. 12

11. Determine the distance X (Fig. 13).

12. Calculate the distance L (Fig. 14).

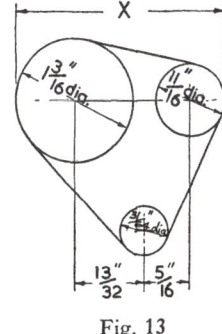

Fig. 13

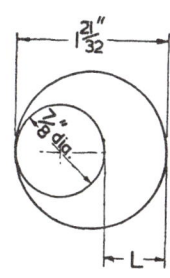

Fig. 14

13. Calculate the radius R (Fig. 15).

14. Determine the distance X (Fig. 16).

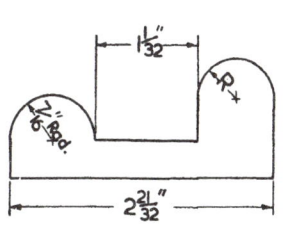

Fig. 15

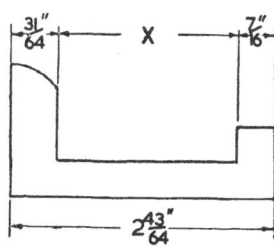

Fig. 16

15. Calculate the length L of the template (Fig. 17).

16. Calculate the dimension X (Fig. 18).

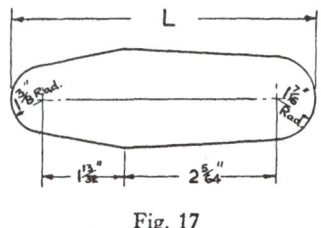

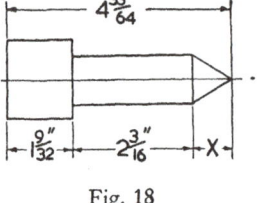

Fig. 17   Fig. 18

17. In the template shown calculate the dimensions a, b and c (Fig. 19).

18. Determine the dimensions A, B and C in the template (Fig. 20).

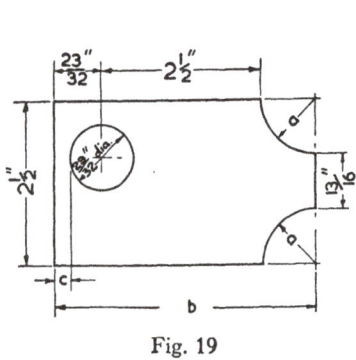

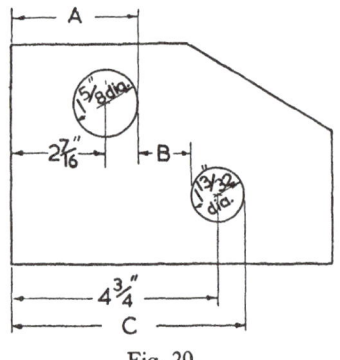

Fig. 19   Fig. 20

19. For the V-block calculate the height h (Fig. 21).

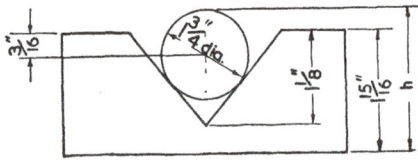

Fig. 21

20. Two test plugs of diameter 1 in. and $2\frac{1}{4}$ in. respectively are fitted into a slot in a bar of steel as shown (Fig. 22). Calculate the dimension d.

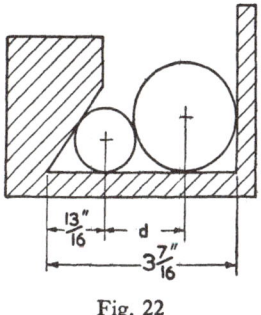

Fig. 22

Exercise 3. Fractions: Multiplication and Division

1. Find the area of a rectangular plate of length $4\frac{4}{5}$ in. and breadth $2\frac{3}{16}$ in.

2. Find the volume and the weight of a metal block of length $3\frac{3}{4}$ in., breadth $3\frac{1}{5}$ in. and height $\frac{3}{4}$ in. if 1 cu.in. of the metal weighs $\frac{1}{4}$ lb.

3. A block of length $2\frac{3}{4}$ in., breadth $2\frac{1}{9}$ in. and height $2\frac{1}{22}$ in. is made of metal of density $\frac{1}{5}$ lb. per cu.in. What is its weight?

4. Find the circumference of a circle of diameter $\frac{13}{32}$ in., given $C = \pi d$ and $\pi = 3\frac{1}{7}$.

5. Calculate the weight of a bar $3\frac{1}{18}$ in. long and $\frac{3}{5}$ in. diameter if the material weighs $\frac{2}{11}$ lb. per cu.in. $\left(\textit{Note.} \text{ Volume of bar} = \dfrac{\pi d^2 l}{4}.\right)$

6. An ingot is cast in the form of a triangular prism $2\frac{2}{3}$ in. long. The base of the triangular end is $\frac{13}{34}$ in. and the altitude is $\frac{17}{32}$ in. Find the volume of the ingot. (*Note.* Volume = area of the end × length.)

7. A hole of diameter $3\frac{1}{2}$ in. is bored in a rectangular plate whose length is $5\frac{1}{4}$ in. and breadth $4\frac{1}{3}$ in. What is the weight of the plate if 1 sq.in. of the material weighs $\frac{1}{3}$ lb.?

8. A wedge has a taper of $\frac{3}{8}$ in. per ft. What is the taper if the wedge is 7 in. long?

9. A wire $2\frac{5}{8}$ in. long is cut into four equal lengths, and each part is then cut into three equal lengths. What is the length of the small division neglecting cutting allowance?

10. The area of a rectangular plate is $4\frac{1}{2}$ sq.in. If its length is $3\frac{3}{8}$ in. what is its breadth?

11. A rectangular block is $3\frac{1}{3}$ in. long and $2\frac{3}{4}$ in. wide. What is the depth if its volume is $20\frac{1}{6}$ cu.in.?

12. How many pins $1\frac{5}{32}$ in. long can be cut from a bar $6\frac{15}{16}$ in. long?

13. If the total area of the metal plate (Fig. 23) is $14\frac{1}{8}$ sq.in., determine the dimension x.

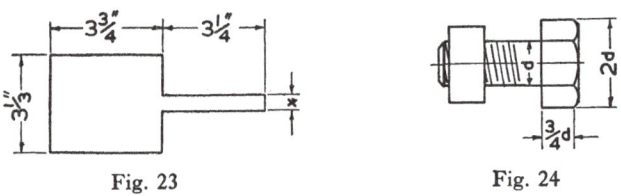

Fig. 23   Fig. 24

14. In Fig. 24 are shown the proportions of a nut and bolt. Calculate the dimensions for bolts of $\frac{3}{4}$ in. and $\frac{7}{8}$ in. diameter.

15. If $F = 1\frac{1}{2}d + \frac{1}{8}$, calculate F when $d = \frac{1}{2}$ in., $\frac{3}{8}$ in. and $1\frac{1}{2}$ in. for the bolt (Fig. 25).

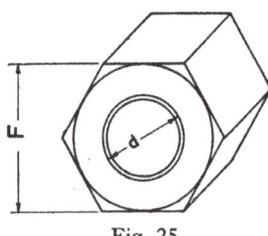

Fig. 25

16. Calculate the weight of a metal ball of diameter $3\frac{1}{2}$ in. if 1 cu.in. of metal weighs $\frac{2}{7}$ lb. (Take $\pi = 3\frac{1}{7}$.)

Exercise 4. Fractions: Miscellaneous Problems

1. A mixture for concrete is made up of 3 parts of 'three-quarter down', 5 parts sand and 2 parts cement. How many pounds of each are in 1 ton of the mixture?

2. A rectangular tank 3 ft. 4 in. long and 2 ft. 6 in. wide contains water to a depth of 2 ft. 3 in. If 1 cu.in. of water weighs $\frac{3}{5}$ oz. (approx.), find the weight of water in the tank in lb.

3. A rectangular block of stone is 5 ft. 4 in. long, 2 ft. 3 in. wide and 3 ft. 2 in. high. If it weighs 3 tons, find the weight of 1 cu.ft. of stone in lb.

4. A cylindrical drum of $10\frac{1}{2}$ in. diameter holds $1\frac{1}{2}$ gallons of oil. Taking $\pi = 3\frac{1}{7}$ and 1 gallon as 288 cu.in., find the depth of oil in the drum to the nearest inch.

5. A machine tool costing 4s. 9d. to produce is sold for 6s. 4d. Express the cost of production as a fraction of the selling price.

6. A train (including the engine) weighs 405 tons. If the engine weighs $\frac{2}{9}$ of the whole find the weight of the engine.

7. A locomotive and tender of B 17 Class 'Sandringham' weighs 129 tons 15 cwt. The engine weighs 77 tons 5 cwt. Express the weight of the engine as a fraction of the total weight.

8. An iron bridge spans a cutting 26 ft. wide. The bridge overlaps the supports one-ninth of its length at one end, and one-sixth of its length at the other end. How long is the bridge?

9. A owns $\frac{3}{7}$ of a building site, B owns $\frac{2}{5}$ and C the remainder which is worth £144. What are A's and B's shares worth?

10. Three partners, A, B and C, buy scrap metal. A buys $\frac{1}{3}$, B buys $\frac{1}{5}$, and C the remainder which weighs 1 ton 8 cwt. What was the total weight bought. If the scrap was melted and recast into ingots each weighing $1\frac{1}{4}$ cwt., how many could be produced assuming no waste?

11. If 1000 metres are taken as being equivalent to $\frac{5}{8}$ mile, express $3\frac{1}{2}$ miles in metres and 750 metres in miles.

12. How many bolts $3\frac{1}{2}$ in. long can be cut from an iron bar 5 ft. 11 in. long? What length remains assuming no waste in cutting?

13. Three iron bars cut from a rod are of length $3\frac{7}{8}$ ft., $4\frac{3}{4}$ ft. and $2\frac{5}{16}$ ft. respectively. The total weight of the bars is $52\frac{1}{2}$ lb. Find the weight per foot run of the rod.

14. Determine the distance L (Fig. 26).

15. Determine the distance X (Fig. 27).

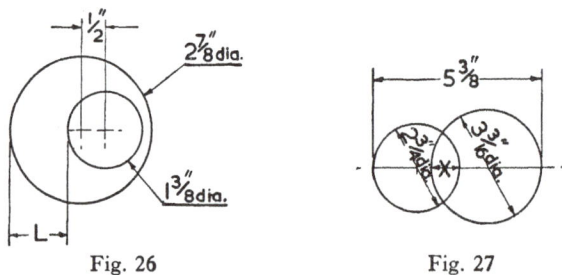

Fig. 26   Fig. 27

Exercise 5. Decimals: Addition and Subtraction

1. Find the combined heating surface of the smoke tubes, firebox, and superheater of the following locomotives:

(*a*) 7P Class. Princess Coronation. 4–6–2:

Smoke tubes. Large and small	2577·0 sq.ft.
Firebox	230·5 sq.ft.
Superheater	856·0 sq.ft.

(*b*) 5 Class. Mixed Traffic. 4–6–0:

Smoke tubes. Large and small	1478·7 sq.ft.
Firebox	171·3 sq.ft.
Superheater	359·3 sq.ft.

(c) 4P Class Compound. 4–4–0:

        Smoke tubes. Large and small   1169·0 sq.ft.

        Firebox   147·3 sq.ft.

        Superheater   290·7 sq.ft.

(d) 4P Class. Parallel Boiler. 2–6–4 T:

        Smoke tubes. Large and small   1082·5 sq.ft.

        Firebox   137·5 sq.ft.

        Superheater   266·25 sq.ft.

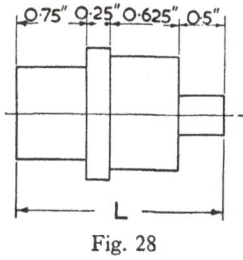

Fig. 28

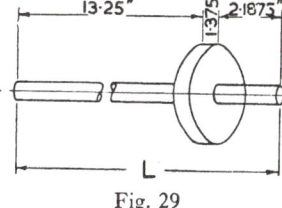

Fig. 29

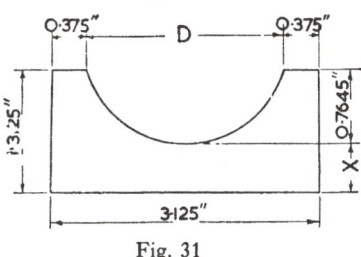

Fig. 30   Fig. 31

2. In a 2P Class 4–4–0 locomotive the combined heating surface is 1410·2 sq.ft. That of the firebox is 123·8 sq.ft. and of the tubes 1033·7 sq.ft. What is the heating surface of the superheater?

3. In a locomotive of 3P Class Taper Boiler, the heating surface of the tubes is 996·4 sq.ft. and that of the superheater 145 sq.ft. If the combined heating surface is 1252·6 sq.ft. what is that of the firebox?

4. Determine the length L of the pin (Fig. 28).

5. Calculate the length L (Fig. 29).

6. Determine the distances D and d (Fig. 30).

7. Determine the dimensions D and X (Fig. 31).

8. Determine the dimension L (Fig. 32).

9. Determine the distance X (Fig. 33).

10. Calculate the lengths L and D (Fig. 34).

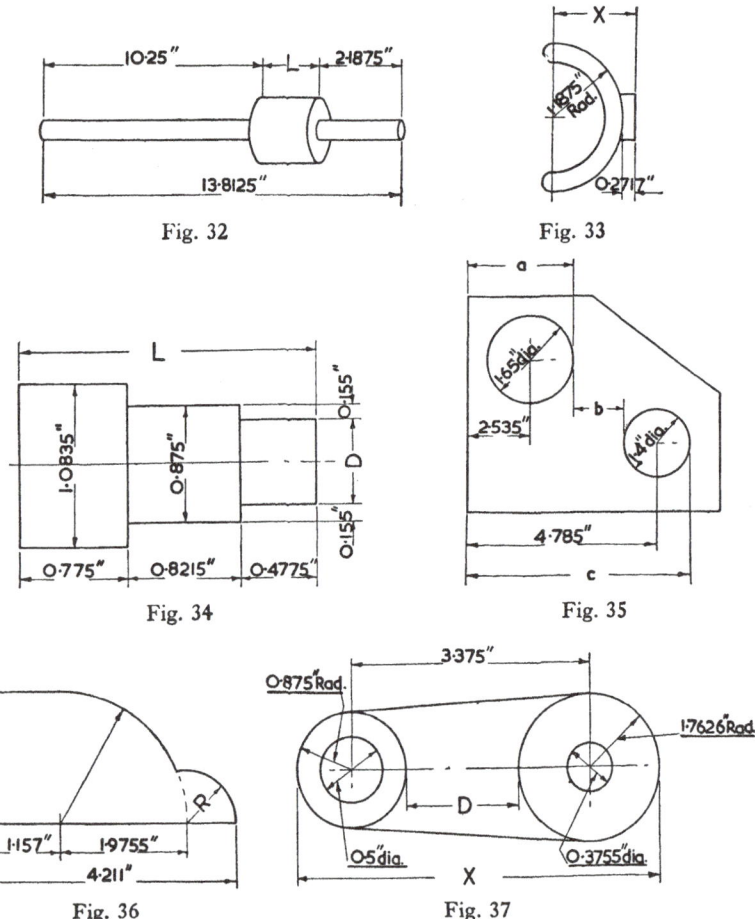

Fig. 32

Fig. 33

Fig. 34

Fig. 35

Fig. 36

Fig. 37

11. For the template calculate the dimensions a, b and c (Fig. 35).

12. Determine the radius R (Fig. 36).

13. Calculate the dimensions X and D (Fig. 37).

Exercise 6. Decimals: Multiplication and Division

1. Evaluate correct to four decimal places:

(a) $52 \cdot 03 \times 0 \cdot 02019$.   (b) $420 \cdot 318 \times 0 \cdot 23175$.

(c) $1 \cdot 4269 \div 58 \cdot 2$.   (d) $0 \cdot 00178478 \div 0 \cdot 0642$.

2. Find the area of a rectangular plate of length 6·4 in. and breadth 2·1875 in.

3. Calculate the volume of a rectangular block of length 3·57 in. breadth 2·85 in. and thickness 1·2 in.

4. Find the diameter of a wheel of circumference 2·9516 ft., given $C = \pi d$ and $\pi = 3 \cdot 142$, giving the answer correct to two decimal places.

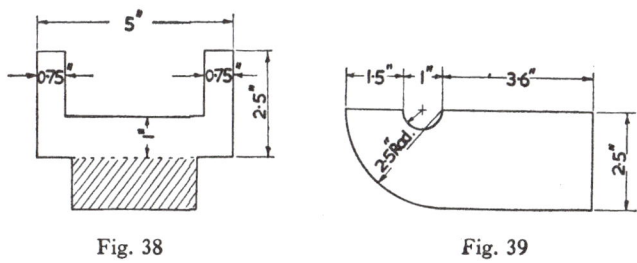

Fig. 38   Fig. 39

5. How many times will a locomotive driving wheel of diameter 5·25 ft. turn in travelling 1·375 miles?

6. A hole of diameter 1·75 in. is bored in a rectangular plate of dimensions 2·75 in. × 2·25 in. Calculate the area remaining. (Take $\pi = 3\frac{1}{7}$.)

7. If the total area of the figure is 11·5 sq.in., find the area of the shaded portion (Fig. 38).

8. Calculate the perimeter and area of Fig. 39, giving your answers correct to one decimal place.

9. Calculate the dimension X if the total area is 29·761 sq.in., giving your answer correct to two places of decimals (Fig. 40).

10. Calculate the area of the shaded portion in Fig. 41 ($\pi=3\frac{1}{7}$.)

11. Determine the perimeter in Fig. 42. ($\pi=3\frac{1}{7}$.)

12. Calculate the area shaded in Fig. 43. ($\pi=3\frac{1}{7}$.)

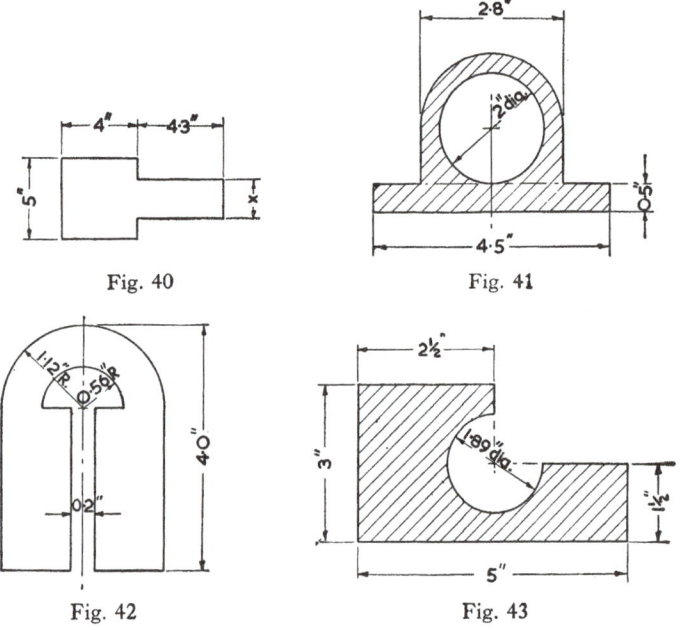

Fig. 40   Fig. 41

Fig. 42   Fig. 43

13. Given the volume V of a cylinder as $V=\dfrac{\pi d^2 L}{4}$, calculate the length L of a cylinder of volume 170·4 cu.in. whose diameter is 2 in. (Take $\pi=3\cdot14$.)

14. An optical formula gives $\dfrac{1}{f}=\dfrac{1}{32}+\dfrac{1}{20\cdot5}$. Calculate the value of f correct to two decimal places.

15. An electrical formula gives $\dfrac{1}{R}=\dfrac{1}{2\cdot5}+\dfrac{1}{4}+\dfrac{1}{5}$. Calculate the value of R correct to three decimal places.

Exercise 7. Decimals: Conversion to Engineering Fractions

1. Find the dimensions of the Snap Head Rivet (Fig. 44), giving the results as fractions to the nearest $\frac{1}{16}$ in. for the cases when the diameter is (a) 0·75 in., (b) 0·875 in., (c) 1·125 in.

2. Find to the nearest $\frac{1}{32}$ in. the dimensions of the Pan Head Rivet (Fig. 45) when the diameter is (a) 1 in., (b) 0·875 in., (c) 1·125 in.

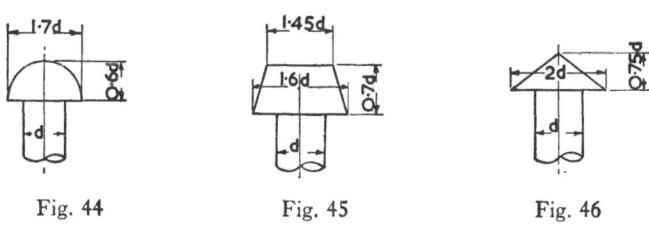

Fig. 44   Fig. 45   Fig. 46

3. Find to the nearest $\frac{1}{64}$ in. the dimensions of the Conical Head Rivet (Fig. 46) when the diameter is (a) 0·75 in., (b) 0·625 in., (c) 1·125 in.

4. The pitch p of the Square Thread shown in Fig. 47 is given by $p = 0·16d + 0·08$. Find the pitch to the nearest $\frac{1}{64}$ in. when d is (a) 0·5 in., (b) 0·875 in., (c) 0·625 in.

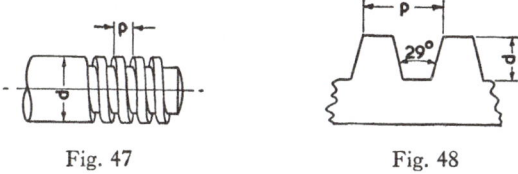

Fig. 47   Fig. 48

5. The depth d of an Acme Thread (Fig. 48) is given by $d = \frac{1}{2}p + 0·01$. Find the depth to the nearest $\frac{1}{64}$ in. when the pitch p is (a) $\frac{3}{16}$ in., (b) $\frac{7}{32}$ in., (c) $\frac{5}{16}$ in.

6. Find to the nearest $\frac{1}{64}$ in. the sizes of washers for bolts of diameter (a) 1 in., (b) $\frac{7}{8}$ in., (c) $1\frac{5}{8}$ in. (Fig. 49), if $D = 2·3d$ and $t = 0·15D$.

7. Convert the dimensions of the portion of the shaft shown in Fig. 50 to the nearest $\frac{1}{64}$ in. when $d = 2 \cdot 625$ in.

8. The depth d of a wedge of length L is given by the formula $d = 0 \cdot 27L + 0 \cdot 17$. Find the depth to the nearest $\frac{1}{32}$ in. when $L = 1 \cdot 35$ in.

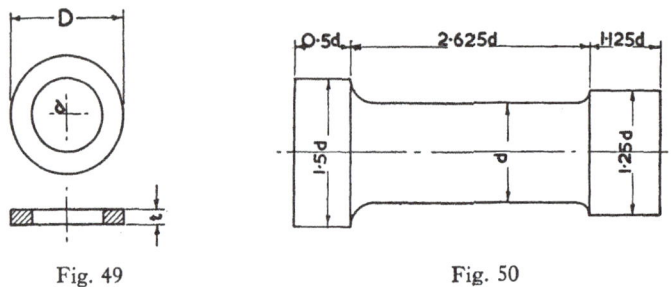

Fig. 49   Fig. 50

9. Calculate to the nearest $\frac{1}{64}$ in. the circumference of a wheel of diameter $1 \cdot 3$ ft. if $\pi = 3 \cdot 14$.

10. What length of angle iron is needed to reinforce all the edges of a closed box $2 \cdot 75$ in. long, $1 \cdot 67$ in. wide and $1 \cdot 28$ in. deep, giving your answer to the nearest $\frac{1}{16}$ in.?

Exercise 8. Spacing of Rivets and Holes

1. If 17 rivets are equally spaced along the side of a tank, the distance between their centres being 3 in., what is the distance between the end rivets?

2. How many rivets, equally spaced 6 in. between their centres, will go round a boiler of circumference 22 ft.?

3. Along a buffer beam are equally spaced 33 rivets, the distance between the centres being 3 in. The end rivets are each 4 in. from their respective ends of the beam. What is the length of the buffer beam?

4. A boiler of diameter 5 ft. 10 in. and length 16 ft. 8 in. is riveted once along its length, and round the circumference near each end. The rivet centres are equally spaced 4 in. apart, and the end rivets are 4 in. from the ends of the boiler. How many rivets are needed. (Take $\pi = 3\frac{1}{7}$.)

5. If 44 holes are equally spaced with centres $4\frac{1}{2}$ in. apart round the circumference of a boiler, what is its diameter?

6. How many quarter-mile posts are there marking a distance of $3\frac{1}{2}$ miles?

7. The end-section of a water tank is 4 ft. 6 in. square. Rivet holes with centres 3 in. apart are drilled round the section with centres 3 in. from the edges of the section. How many holes are made?

8. Holes are to be drilled in a circular plate along three circles of pitch diameters 3 ft. 6 in., 1 ft. 9 in. and $10\frac{1}{2}$ in. respectively. The outer set of holes are to have their centres 4 in. apart, and the middle set 3 in. apart. If the centres of the inner set are 1 in. apart how many holes are drilled?

9. Along a buffer beam of length 8 ft. $11\frac{1}{4}$ in. are drilled 33 rivet holes of $\frac{3}{8}$ in. diameter. The holes are equally spaced and the end holes are each $3\frac{5}{8}$ in. from the ends of the beam. How are the holes spaced?

10. Through a plate $22\frac{3}{8}$ in. long, 12 holes of square section are cut in a line, the clearances being $1\frac{1}{8}$ in. What is the width of each hole if the clearance at each end is to be 2 in.?

Exercise 9. Algebra: Generalized Addition

Perform the following additions:

1. $+5$ $+9$	**2.** $+6$ -8	**3.** -7 $+4$	**4.** -9 -7	**5.** $+6$ -2
6. -12 -10	**7.** -17 $+16$	**8.** -15 $+21$	**9.** $+27$ $+5$	**10.** -17 $+9$

11. $+7$	**12.** -6	**13.** $+14$	**14.** -7	**15.** $+4$
-4	$+7$	-3	-5	$+7$
$+3$	-9	-11	$+12$	-12

16. $7+(+9)+(+4)$.   **17.** $6+(-8)+4$.

18. $21+(-6)+(-4)$.   **19.** $(-27)+(-5)+(-4)$.

20. $(-4)+(-8)+12$.

Exercise 10. Algebra: Generalized Subtraction

Perform the following subtractions:

1. $+6$	**2.** $+8$	**3.** $+6$	**4.** -6	**5.** $+19$
$+2$	-2	$+12$	$+7$	-3

6. -27	**7.** -18	**8.** $+9$	**9.** -34	**10.** -59
$+4$	-6	-10	$+27$	-41

11. $10-(+6)$.   **12.** $10-(-6)$.

13. $7-(+9)$.   **14.** $9-(-12)$.

15. $(-6)-(+14)$.   **16.** $(-8)-(-14)$.

17. $40-(+30)$.   **18.** $37-(-3)$.

19. $(-15)-(+4)$.   **20.** $(-17)-(-17)$.

Exercise 11. Algebra: Generalized Multiplication

Perform the following multiplications:

1. 5	**2.** 7	**3.** -6	**4.** -8	**5.** -7
3	-3	4	-2	2

6. -4	**7.** -8	**8.** 5	**9.** -1	**10.** -1
7	-9	-8	12	-1

11. $(5)(7).$ **12.** $(7)(-3).$

13. $(-9)(6).$ **14.** $(-6)(-5).$

15. $(2)(-3)(-4).$ **16.** $(-5)(-2)(-8).$

17. $(3)(-5)(-3)(-2).$ **18.** $(-1)(-2)(-3).$

19. $(-4)(-1)(-2)(-3).$ **20.** $(-1)(-1)(-1)(-1)(-1).$

Exercise 12. Algebra: Generalized Division

Perform the following divisions:

1. $\dfrac{12}{4}$ **2.** $\dfrac{14}{-2}$ **3.** $\dfrac{-18}{6}$ **4.** $\dfrac{-15}{-3}$

5. $\dfrac{54}{-9}$ **6.** $(20)\div(-4).$ **7.** $(-24)\div6.$

8. $(-45)\div(-15).$ **9.** $(-48)\div16.$ **10.** $\dfrac{(-17)}{(-17)}.$

11. $\dfrac{(3)(2)}{(5)(7)}.$ **12.** $\dfrac{(3)(-5)}{(4)(7)}.$ **13.** $\dfrac{(-5)(-8)}{(7)(9)}.$

14. $\dfrac{(3)(5)}{(4)(-7)}.$ **15.** $\dfrac{(7)(9)}{(-10)(-10)}.$ **16.** $\dfrac{(-6)(-4)}{(-7)(-11)}.$

17. $\dfrac{(-5)(-8)(-3)}{(-6)(10)(4)}.$ **18.** $\dfrac{0{\cdot}5\times200\times21}{4{\cdot}2}.$

19. $\dfrac{(-32)(10)(10)}{2}.$ **20.** $\dfrac{300\times3\times110\times2000}{33000}.$

Exercise 13. Algebra: Symbols, Addition

Perform the following additions:

1. $\begin{array}{c} 5x \\ \underline{3x} \end{array}$ **2.** $\begin{array}{c} 3a \\ \underline{7a} \end{array}$ **3.** $\begin{array}{c} 6b \\ \underline{-4b} \end{array}$ **4.** $\begin{array}{c} -5c \\ \underline{8c} \end{array}$ **5.** $\begin{array}{c} -8x \\ \underline{-12x} \end{array}$

6. $4x+2x+3x+7x.$ **7.** $4a-2a+9a-a-2a.$

8. $6b - 4b - 3b + 5b + 3b - 2b.$ **9.** $2x + 3y - 5x - 4y - 8x - 5y.$

10. $2a + 4b + 3c + 5a - 2b + 6c - 3a - b - c.$

11. $5a + 3b - 4c$ **12.** $3a^2 + 2a - 6$
        $3a - b + 7c$ $4a^2 - 7a + 2$
    $-4a - 3b + 2c$ $-2a^2 - 3a - 7$

13. $a^2b + ab^2 - ab$ **14.** $4u - 3v + 2w$
       $4a^2b \qquad + 7ab$ $2u - v - 7w$
     $3a^2b + 4ab^2 + 3ab$ $3u - 2v$
   $-7a^2b - 2ab^2 - 4ab$ $4v - 3w$

15. Find the sum of $2a + b - 2c,\ 3a + 2b + c,\ a - 4b + 5c.$

16. Find the sum of $7x^3 - 3x^2 + 2x^2 - 4x + 2x^3 - 6 + 5x + 5x^2.$

17. Find the perimeter of a triangle whose sides are $a + 5b,\ 2a + 3b$ and $3a - b.$

18. What is the perimeter of a plate whose sides are $7a - 3b - 2c,$ $3a - b,\ a + 2b$ and $5a - 4b + c,$ the dimensions being in feet?

19. The distances between the centres of six rivets placed in a line are $2x,\ 2x + b,\ 3x - 2b,\ b$ and $x + 4b$ in. respectively. Find the distance between the extreme end rivets.

20. If $P = x + y,\ Q = 3x - 2y$ and $R = 5y - 7x,$ find the value of $P + Q + R$ in terms of x and $y.$

Exercise 14. Algebra: Symbols, Subtraction

Perform the following subtractions:

1. $8a$ **2.** $3ab$ **3.** $-4x^2y^2$ **4.** $-5ax$
    $5a$ $-2ab$ $-5x^2y^2$ ax

5. $3a + 5b$ **6.** $6a - 2b + 5c$ **7.** $10a^2 - 7a - 5$
    $2a + 7b$ $8a + b - 7c$ $-8a^2 - 5a + 8$

8. $3a^2b - 7ab + 6ab^2$
$\underline{3a^2b - 8ab - 6ab^2}$

9. $5u - 3v + 6w$
$\underline{-3u - 3v + 7w}$

10. $3a - 5b$
$\underline{6a + 2b - 3c}$

11. From $3a + 5b + 4c$ take $4a - 5b - 5c$.

12. From $5a - 7b - 3c$ take $2a + 4b - 7c$.

13. From $a^3 - 2a^2b + 3ab^2 - b^3$ take $3a^2b - ab^3$.

14. Take $3a - 5b$ from the sum of $5a - 7b + 3c$ and $-4a - 4b + 11c$.

15. Find the sum and difference between
$$7a^2 - 7a + 6 \text{ and } -9a^2 - 8a + 4.$$

16. From $4u - 12v$ take $-2u + 5k + 3v$.

17. From a piece of timber x ft. long a piece $(a + b)$ in. long is cut off. What length in inches remains?

18. If a man starts on a journey of S miles and walks for y hours at x miles per hour, how far has he still to go?

19. From the square of x take the square of y, and subtract $3xy + y^2$ from the result.

20. A man walks $(3x - y)$ miles due north from a point A, and then walks $(5x + 3y)$ miles due south. What is his position now with regard to the point A?

Exercise 15. Algebra: Symbols, Multiplication

Perform the following multiplications:

1. $3x \cdot 2x$.   **2.** $5x \cdot 3x^2$.   **3.** $6a^3 \cdot 2a^2$.   **4.** $4m^4 \cdot m$.

5. $6a^2b \cdot 3ab^2$.   **6.** $2a^3b^3c^3 \cdot 3a^2b^3c$.   **7.** $6u^2v \cdot 3uv^2$.

8. $(-2ab)(3ab)(4ab)$.   **9.** $(-2a)(4a^2)(-1)$.

10. $(-3x^2y)\,(-4xy)\,(-2xy^2)$.   11. $a(a^2-2a-3)$.

12. $x(2x^3-3x+1)$.   13. $2y(2y^2-3y+6)$.

14. $c^2(c^3-c^2-c+1)$.   15. $3a^2b(3a^2b-2ab^2)$.

Exercise 16. Algebra: Polynomials and Problems

Perform the following multiplications:

1. $(x+2)\,(x+4)$.   2. $(a+4)\,(a-3)$.

3. $(a-5)\,(a-3)$.   4. $(a-7)\,(a+10)$.

5. $(2a-5)\,(3a+2)$.   6. $(3x-7)\,(5x-4)$.

7. $(4c-6)\,(c-5)$.   8. $(x+2)\,(x+2)$.

9. $(x-4)\,(x-4)$.   10. $(2x-3)\,(2x^2-3x+4)$.

11. $(2x+3y)\,(3x+4y)$.   12. $(3x-2y)\,(3x+2y)$.

13. $2(x-3y)\,(x+y)$.   14. $\pi(R-r)\,(R+r)$.

15. $\dfrac{\pi}{4}\,(D-d)\,(D+d)$.

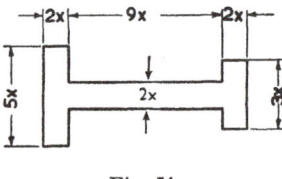

Fig. 51

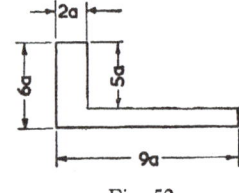

Fig. 52

16. Find the area of a rectangular plate of length $5a$ ft. and breadth $3a$ ft.

17. What is the area of a triangle whose base is $6x$ in. and altitude $3x$ in.?

18. Find the area of the section shown (Fig 51).

19. What is the area of the cross-section of a railway embankment which is $12x$ ft. wide at the bottom, and $8x$ ft. wide at the top if it is $6x$ ft. high?

20. Find the area of the L section shown in Fig. 52.

21. A rectangular tool box is $(x+3)$ in. long, $(x+2)$ in. wide and $(x+1)$ in. deep. What is its volume?

Exercise 17. Algebra: Symbols, Division

Perform the following divisions:

1. $\dfrac{x^5}{x^2}$.   **2.** $\dfrac{-a^6}{a}$.   **3.** $\dfrac{x^5}{-x}$.   **4.** $\dfrac{-a^4}{-a^3}$.   **5.** $\dfrac{-15c^5}{3c^3}$.

6. $\dfrac{16x^4}{-2x^3}$.   **7.** $\dfrac{-21a^7}{-3a}$.   **8.** $\dfrac{-35d^5}{5d^2}$.   **9.** $\dfrac{3d^4}{d^4}$.   **10.** $\dfrac{4a^7}{8a^3}$.

11. $\dfrac{x^2}{x^3}$.   **12.** $\dfrac{-a^3}{a^7}$.   **13.** $\dfrac{b^5}{-b^8}$.   **14.** $\dfrac{-c^2}{-c^4}$.   **15.** $\dfrac{x^4y^3}{x^2y}$.

16. $\dfrac{-a^5b^6}{a^5b^4}$.   **17.** $\dfrac{-c^3d^5}{-cd^2}$.   **18.** $\dfrac{12a^4b^5c^6}{3a.b.c}$.

19. $\dfrac{25xyz}{5x^2y^3z^4}$.   **20.** $\dfrac{3a^2.5b^3.21c^5}{15a.7b^5.c^2}$.

Exercise 18. Algebra: Simple Equations

Solve the following equations and check the results:

1. $2x=8$.   **2.** $3x=9$.   **3.** $5x=15$.   **4.** $7a=21$.

5. $8b=32$.   **6.** $3x=3$.   **7.** $4a=0$.   **8.** $4b=-12$.

9. $-5c=20$.   **10.** $-6x=-24$.   **11.** $2x=1$.

12. $3a=2$.   **13.** $7b=5$.   **14.** $12c=3$.

15. $6x=-2$.   **16.** $-8a=4$.   **17.** $-5b=-1$.

18. $-14c = -7.$ **19.** $6x = -6.$ **20.** $-8x = 0.$

21. $2x = \frac{1}{2}.$ **22.** $5x = \frac{1}{3}.$ **23.** $2x = \frac{3}{4}.$

24. $4x = \frac{4}{5}.$ **25.** $7x = -\frac{3}{5}.$ **26.** $-6x = \frac{3}{8}.$

27. $-7x = -\frac{7}{8}.$ **28.** $2x = \frac{5}{3}.$ **29.** $5x = -\frac{15}{16}.$

30. $-9x = -\frac{3}{32}.$ **31.** $\frac{2}{3}x = 1.$ **32.** $\frac{3}{4}x = 6.$

33. $\frac{4}{3}a = \frac{6}{7}.$ **34.** $\frac{2}{3}b = \frac{1}{2}.$ **35.** $\frac{2}{3}c = \frac{3}{4}.$

36. $\frac{3}{4}d = \frac{5}{3}.$ **37.** $\frac{1}{2}x = -\frac{1}{4}.$ **38.** $-\frac{1}{3}x = \frac{1}{12}.$

39. $-\frac{1}{5}x = -\frac{1}{15}.$ **40.** $\frac{4}{5}a = \frac{4}{3}.$

Exercise 19. Algebra: Equations, Fractions

Solve the following equations:

1. $2x + 4x = 12.$ **2.** $3a + 5a = 24.$

3. $6x - 2x = 8.$ **4.** $12b - 9b = 6.$

5. $-4x + 6x = 14.$ **6.** $8a - 3a = 12 + a.$

7. $(9b - 4b) = (21 - 6).$ **8.** $5x + 2x = 15 - 11.$

9. $6x - 3 = 15.$ **10.** $4a - 5 = 11.$

11. $2b + 14 = 24.$ **12.** $6x - 7 = 23.$

13. $9x = 22 - 2x.$ **14.** $5x = 14 - 2x.$

15. $6x - 4 = 5x + 2.$ **16.** $7x + 3 = 8x - 2.$

17. $3x + 3 = 4x - 2.$ **18.** $2x - 27 = -15 - 4x.$

19. $17 - 7x = 23 - 10x.$ **20.** $-4y - 8 = -7y + 3.$

21. $-4y - 8 = -7y + 31.$ **22.** $\frac{1}{2}x + \frac{1}{3}x = 10.$

23. $\frac{1}{2}x - \frac{1}{3}x = 1.$

24. $\frac{1}{3}x + \frac{1}{4}x = 2.$

25. $\frac{1}{3}x - \frac{1}{6}x = 3.$

26. $\frac{2}{3}x + \frac{3}{4}x = 4.$

27. $x = \frac{1}{3}x + 5.$

28. $a = 5 + \frac{3}{4}a.$

29. $\frac{1}{2}b + \frac{1}{3}b = b - 6.$

30. $\frac{4}{5} = \frac{2}{3}x.$

31. $\frac{6}{7x} = \frac{9}{4}.$

32. $\frac{9}{5} = \frac{8}{5x}.$

33. $\frac{10}{3} = \frac{5}{2x}.$

34. $\frac{6}{5} = \frac{6}{5x}.$

35. $\frac{2}{7} = \frac{3}{x}.$

Exercise 20. Algebra: Equations, Fractions, Brackets

Solve the equations:

1. $8(1 + x) = 5x - 4.$

2. $7(x - 2) = 5(x - 4).$

3. $4(2x - 3) = 3(5x - 6).$

4. $12 - 5(x + 2) = 2(x - 1).$

5. $5(x - 7) = 9 - 3(x - 4).$

6. $3(2x - 5) - 10(x - 3) = 3.$

7. $4(x + 2) - (x - 4) = x - 2.$

8. $8 - \{3x - (4x - 5) - 1\} = 0.$

9. $3(x + 2) - 4(x - 1) = x - 2.$

10. $10x - 2[3x - 3\{2(x - 1)\} + 2] = 16.$

11. $\frac{2x + 1}{3} = \frac{3x - 1}{2}.$

12. $\frac{2x + 6}{5} = \frac{3x - 2}{3}.$

13. $\frac{5x - 11}{5} + \frac{9 - 4x}{7} = 0.$

14. $\frac{4a + 2}{5} = \frac{3(1 - a)}{3}.$

15. $\frac{2b + 5}{2} = \frac{3 - 2b}{5}.$

16. $\frac{2c - 1}{3} - \frac{3c - 2}{4} = \frac{5c - 4}{6} - \frac{7c + 6}{12}.$

17. $\frac{1}{2}(x + 1) + \frac{1}{3}(x + 3) = \frac{1}{5}(x + 4) + 2.$

18. $\frac{x + 4}{3x} = \frac{3}{5}.$

19. $\frac{3x - 1}{2x} = \frac{3}{4}.$

20. $\frac{3}{x - 6} = \frac{2}{x + 5}.$

Exercise 21. Algebra: Harder Equations

Solve the equations:

1. $(x+1)(x+2)=x^2+8.$　　**2.** $(x+1)(x+3)=(x+4)(x-1).$

3. $(x-1)(x+1)=(x+1)^2.$　　**4.** $2(x-1)(x+1)=(x-2)(2x+3).$

5. $(x+2)(x+3)+(x+1)(x+4)=2x^2-5.$

6. $4x(2x+1)-10x=8(x+5)(x-6)+300.$

7. $(x+6)(x-2)-(x+2)(x+3)=12-7x.$

8. $(3x+4)(4x-2)-(5x-2)(x+1)=(7x-1)(x-1)+8.$

9. $4(x+5)^2-3(x-5)^2=10+(x-1)^2.$

10. $2(x+4)(x-4)-(x-4)^2=(x+4)^2-13x+1.$

Exercise 22. Algebra: Simple Literal Equations

Solve for x, y or z:

1. $2x=4a.$　　**2.** $3x=9b^2.$　　**3.** $ax=4a.$　　**4.** $bx=b.$

5. $a+x=b.$　　**6.** $z-c=c.$　　**7.** $y-a+b=c.$　　**8.** $az=b.$

9. $cz+a=d.$　　**10.** $4x=bc-d.$　　**11.** $ay+b=3c-d.$

12. $4x=8bc-4d.$　　**13.** $6x=a+2x.$　　**14.** $\dfrac{y}{2}=a.$

15. $\dfrac{y}{3}=\dfrac{b}{2}.$　　**16.** $\dfrac{3x}{4}=\dfrac{5cd}{8}.$　　**17.** $\dfrac{3a}{5}=\dfrac{1}{x}.$

18. $\dfrac{3x-1}{2}=\dfrac{a-b}{3}.$　　**19.** $\dfrac{ax-1}{3}=\dfrac{ax+3}{4}.$　　**20.** $\dfrac{by-1}{y+2}=\dfrac{a}{b}.$

Exercise 23. Algebra: Problems on Equations

1. An iron bar 19 ft. long is cut into two pieces so that one piece is 3 ft. longer than the other. What are the lengths of the pieces?

2. The perimeter of a rectangle is 44 in. If one of the adjacent sides is 1·8 in. longer than the other what are the dimensions?

3. A triangular plate of sides a, b and c is such that b is twice the length of a, and c is 3 in. longer than a. If the perimeter is 11 in., find the lengths of the sides.

4. A rectangular box with square ends has its length 10 in. longer than its breadth, and the total length of all its edges is 152 in. What is its breadth?

5. Two rectangles of equal area have widths 10 in. and 12 in. The difference in their lengths is 3 in. Find the area of the rectangles.

6. The length of a rectangular lawn is twice its breadth. If a strip 4 ft. wide is cut off all round it, the area is diminished by 440 sq.ft. Find the original length and breadth.

7. A rectangular metal plate is 24 cm. long. A strip 3·5 cm. wide is cut off from one end, and a second strip 2·25 cm. wide is cut off from the other end. The remainder weighs 153·3 gm. Find the width of the plate if 1 sq.cm. of the metal weighs 0·7 gm.

8. The breadth of a rectangular water tank is 4 times its depth, and the length is 5 times its breadth. If it holds 10,000 cu.ft. of water, find its dimensions.

9. If the initial velocity (u), and the final velocity (v) of a particle moving with constant acceleration (f) are known, then the distance travelled (s) can be calculated from the formula $v^2 - u^2 = 2fs$. Find s when $u = 26$, $v = 74$ and $f = 32$.

10. The relation between Centigrade and Fahrenheit temperatures is given by $F = \frac{9}{5}C + 32$. Change 98° F. to Centigrade temperature.

11. Find at what temperature the Fahrenheit thermometer will show double the reading of the Centigrade thermometer.

12. The circumferences of two wheels are 26 in. and 28 in. respectively. In travelling a certain distance one of them makes 30 revolutions more than the other. Find the distance travelled in feet.

13. The current passing through an electrical circuit is given by the formula $I = \dfrac{E-e}{R}$. Find the applied e.m.f. (E) when $I = 100$, $e = 240$ and $R = 0.05$.

14. The resistance of a parallel electrical circuit is given by

$$\frac{1}{R} = \frac{1}{r_1} + \frac{1}{r_2}.$$

Calculate the value of r_1 if $R = 18$ and $r_1 = 3r_2$.

15. A crew which can row at the rate of 5 miles per hour in still water finds that it takes twice as long to row up the river as it does to row down. At what rate does the river flow?

16. A square plate has one side increased by 6 in. and the other decreased by 4 in. without changing its area. What are the dimensions of the plate?

17. A rectangle has one side 20 in. longer than the other. If the longer side were decreased by 26 in. and the shorter side increased by 30 in. the area would be unchanged. What are the dimensions of the original rectangle?

18. A formula connected with the stress (f) in the material of a cylinder is $t = \dfrac{pd}{2f} + \dfrac{1}{4}$. Find f for the case when $p = 80$, $d = 12$ and $t = \frac{5}{8}$.

19. In the formula $s = ut + \frac{1}{2}ft^2$, find f when $s = 450$, $u = 10$, and $t = 5$.

20. If $c = \sqrt{(a^2 + b^2)}$, find the value of b when $c = 13$ and $a = 12$.

Exercise 24. Algebra: Substitution (positive numbers)

If $a=2$, $b=3$, $c=1$, $x=4$, $y=6$, find the value of:

1. $2a$.
2. a^2.
3. $3a$.
4. a^3.
5. $4b$.

6. b^4.
7. $3y^2$.
8. $5c^3$.
9. x^4.
10. a^5.

11. $\dfrac{1}{x^2}$.
12. $\dfrac{b^2}{y^3}$.
13. $\dfrac{b^4}{y^4}$.
14. $\dfrac{a^3}{c^2}$.
15. $\dfrac{a^2}{6b^2}$.

16. $\dfrac{2b^2}{3y^2}$.
17. $\dfrac{6bc}{2ax}$.
18. $\dfrac{cx}{ay}$.
19. $\dfrac{abc}{xy}$.
20. $\dfrac{b^2c^2x^2}{a^3y^2}$.

21. 2^b.
22. 4^a.
23. x^a.
24. $\dfrac{a^b}{x^a}$.
25. $3y^a$.

If $a=1$, $b=2$, $c=3$, $d=0$, $x=4$, $y=5$ find the value of:

26. b^2c.
27. bc^2.
28. cy^3.
29. b^3x.
30. cy^2d.

31. c^2bx^2.
32. a^4b^4c.
33. $3ab^2dx$.
34. $\dfrac{ab^2d}{x^2y}$.
35. $\dfrac{7a^2b^2}{a^4b^4}$.

If $a=2$, $b=3$, $c=4$, $d=1$, $x=6$, $y=10$, $z=0$ find the value of:

36. $3a+2b+c$.
37. $3b+2c-4x$.

38. $6c-2a+3x$.
39. $5x+4y+6z$.

40. $5c-4d+2x$.
41. $2a+3b-3c+4d-5z$.

42. $2ab+3cd-2xy+4cz$.
43. $\dfrac{ab}{xy}-\dfrac{bd}{ay}$.

44. $\dfrac{cd}{a}-\dfrac{ac}{b}$.
45. $\dfrac{2xz}{3bc}+\dfrac{4ad}{6x}$.

If $a=2$, $b=1$, $c=3$, $x=4$, $y=0$ evaluate:

46. $a^2+b^2+c^2$.
47. $x^3-3c^2-4a+5y^2$.

48. $a^3-2bc^2-3c^2x$.
49. $c^4-3ac+a^4$.

50. $a^3-3ab^2+3a^2b-b^3$.

Exercise 25. Algebra: Substitution (negative numbers)

If $a = -1$, $b = -2$, $c = -3$, $d = 0$ find the value of:

1. $3b$.   **2.** b^2.   **3.** c^3.

4. $-3c$.   **5.** $-4b^2$.   **6.** $(-c)^2$.

7. $-b^3$.   **8.** $(-b)^3$.   **9.** $a^2 b^2$.

10. $-a^2 b^2 c^2$.   **11.** $a^2 + b^2 - c^2 + d^2$.   **12.** $a^3 - b^3 + c^3$.

13. $2a + 5b - 3c + 4d$.   **14.** $ab - 3bc - 2ac$.   **15.** $b^2 + c^2 + d^2$.

16. $2b^2 c - 3d + a$.   **17.** $abc + bcd - dab$.   **18.** $\dfrac{ab}{2bc}$.

19. $\dfrac{3cd}{4a^3}$.   **20.** $\dfrac{a^3}{b^3} + \dfrac{a}{bc}$.

Exercise 26. Formula Manipulation and Substitution

Change the 'Subject' in the following formulae and perform the required calculation:

No.	Formula	Subject	Given	Calculate
1	$A = LB$	L	$A = 12$, $B = 3$	L
2	$V = L.B.H$	B	$V = 48$, $L = 6$, $H = 4$	B
3	$P = \dfrac{1}{n}$	n	$P = 12$	n
4	$A = \dfrac{V}{L}$	L	$A = 15$, $V = 125$	L
5	$L = D + 2K$	K	$L = 5$, $D = 4 \cdot 5$	K
6	$d = 2C - D$	C	$d = 1 \cdot 4$, $D = 4 \cdot 1$	C
7	$R = R_1 + R_2 + R_3$	R_3	$R = 14$, $R_1 = 4$, $R_2 = 3$	R_3
8	$v = u + gt$	t	$v = 815$, $u = 15$, $g = 32$	t
9	$V = \pi r^2 h$	r	$V = 5\frac{1}{7}$, $\pi = 3\frac{1}{7}$, $h = 6\frac{6}{11}$	r
10	$V = \dfrac{\pi d^2 l}{4}$	d	$V = 115\frac{1}{2}$, $\pi = 3\frac{1}{7}$, $l = 12$	d

No.	Formula	Subject	Given	Calculate
11	$P = \dfrac{N+n}{2C}$	n	$P=6,\ C=7,\ N=60$	n
12	$C = \dfrac{N-n}{2P}$	n	$N=100,\ C=5,\ P=6$	n
13	$P = \dfrac{S(C-F)}{C}$	C	$P=\frac{1}{5},\ S=\frac{1}{4},\ F=\frac{4}{5}$	C
14	$H = \dfrac{C^2RT}{J}$	C	$H=6000,\ R=336,$ $T=300,\ J=4\cdot2$	C
15	$H = 0\cdot06\ C^2RT$	T	$H=40000,\ C=10,$ $R=600$	T
16	$S = \pi r(r+h)$	h	$S=99,\ \pi=3\frac{1}{7},\ r=3\frac{1}{2}$	h
17	$R = r + art$	r	$R=230,\ a=0\cdot005,\ t=30$	r
18	$I = \dfrac{E}{R}$	R	$I=6,\ E=200$	R
19	$I = \dfrac{E}{R+r}$	r	$E=2\cdot5,\ I=0\cdot05,\ R=30$	r
20	$A = \dfrac{\pi(D^2-d^2)}{4}$	d	$A=2\frac{3}{4},\ D=3\frac{3}{4},\ \pi=3\frac{1}{7}$	d
21	$r = \sqrt{(0\cdot36A)}$	A	$r=4\cdot2$	A
22	$a = \sqrt{(b^2+c^2)}$	b	$a=5,\ c=3$	b
23	$L = c + \sqrt{D}$	D	$L=12,\ c=5$	D
24	$a = b + \sqrt{(b^2+c^2)}$	b	$a=6,\ c=4$	b
25	$T = 2\pi\sqrt{\dfrac{l}{g}}$	g	$T=2\frac{5}{14},\ l=4\frac{1}{2},\ \pi=3\frac{1}{7}$	g
26	$S = \dfrac{n}{2}(a+l)$	n	$S=4995,\ a=11,\ l=100$	n
27	$l = a(n-1)\,d$	d	$l=-10,\ a=6,\ n=7$	d
28	$Ah = 12(P-Q)$	P	$A=\frac{3}{5},\ h=20,\ Q=3$	P
29	$l = ar^2$	r	$l=36,\ a=4$	r
30	$\dfrac{1}{a} + \dfrac{1}{b} = c$	b	$a=2,\ c=\frac{5}{8}$	b

Exercise 27. Harder Formula Manipulation

A

Express x in terms of the other quantities:

1. $4ax + 5 = 2ax + 7$.   2. $ax + 5b = 7b$.

3. $4x + 5b = x + 3(b + c)$.   4. $3x - 2b = 3(b - c - 2x)$.

5. $2x - 3a = 3 + (a - 2x)$.   6. $2(x - 3a) = 3a - 2(b + x)$.

7. $ax - b = c - 2(x + a)$.   8. $a(x - b) = cx + a$.

9. $a(2x + b) - ac = b(2x - a)$.   10. $2(x + a) - a(x + 2) = 2 - (x + a)$

B

If

1. $\dfrac{x}{2} + \dfrac{a}{3} = \dfrac{x}{6} - \dfrac{a}{4}$, find a formula for x.

2. $\dfrac{x}{a} + \dfrac{y}{b} = 2$, express x in terms of a, b and y.

3. $\dfrac{1}{u} + \dfrac{1}{v} = \dfrac{1}{f}$, express v in terms of u and f.

4. $f(M + m) = g(M - m)$, find a formula for m.

5. $t = \dfrac{aW}{c - bW}$, find a formula for W.

6. $C = \frac{5}{9}(F - 32)$, find a formula for F.

7. $m = \dfrac{an^2}{1 - bn^2}$, find a formula for n.

8. $S = \dfrac{(u + v)\, t}{2}$, find a formula for v.

9. $S = ut + \frac{1}{2}ft^2$, find a formula for u.

10. $a = \dfrac{(x - y)}{(x + y)}\, b$, find a formula for y.

Exercise 28. Square Root (Arithmetical)

A

Without the use of tables find the square root of the following:

1. 169.   **2.** 225.   **3.** 484.   **4.** 1089.   **5.** 3969.

6. 8836.   **7.** 189·0625.   **8.** 2905·21   **9.** 1000.   **10.** 9702·25.

B

Calculate the value of the square root of the following, correct to the second place of decimals:

1. 1·57.   **2.** 8·759.   **3.** 0·9375.   **4.** 0·1297.   **5.** 0·0854.

6. 0·125.   **7.** 0·0625.   **8.** 113·75.   **9.** 52·4.   **10.** 800·7.

Exercise 29. Square Root and Squares from Tables

From tables of square roots find (a) the square root, (b) the square of the following:

1. 3·75.   **2.** 17·5.   **3.** 28·76.   **4.** 68·98.

5. 99·79.   **6.** 114·6.   **7.** 47·82.   **8.** 85·37.

9. 325·6.   **10.** 897·3.   **11.** 0·9876.   **12.** 0·3905.

13. 0·1578.   **14.** 0·0857.   **15.** 0·063.   **16.** 0·03125.

17. 0·0075.   **18.** 1·002.   **19.** 0·01.   **20.** 0·9999.

Exercise 30. The Theorem of Pythagoras: Simple Problems

1. In a triangle ABC of sides a, b, c the angle $ACB = 90°$.

(a) If $c = 10$, $b = 8$, determine the length a.

(b) If $a = 7$, $b = 24$, determine the length c.

(c) If $c = 41$, $b = 40$, determine the length a.

(d) If $c = 3\frac{1}{4}$, $a = 1\frac{1}{2}$, determine the length b.

(e) If $a = 2.75$, $b = 4.68$, determine the length c.

(f) If $c = 12.25$, $b = 8.74$, determine the side a.

(g) If $c = 24.76$, $a = 14.87$, determine the side b.

2. In a right-angled triangle the hypotenuse is x in. long. The other two sides are respectively 3 in. and $(x-1)$ in. long. Calculate the value of x.

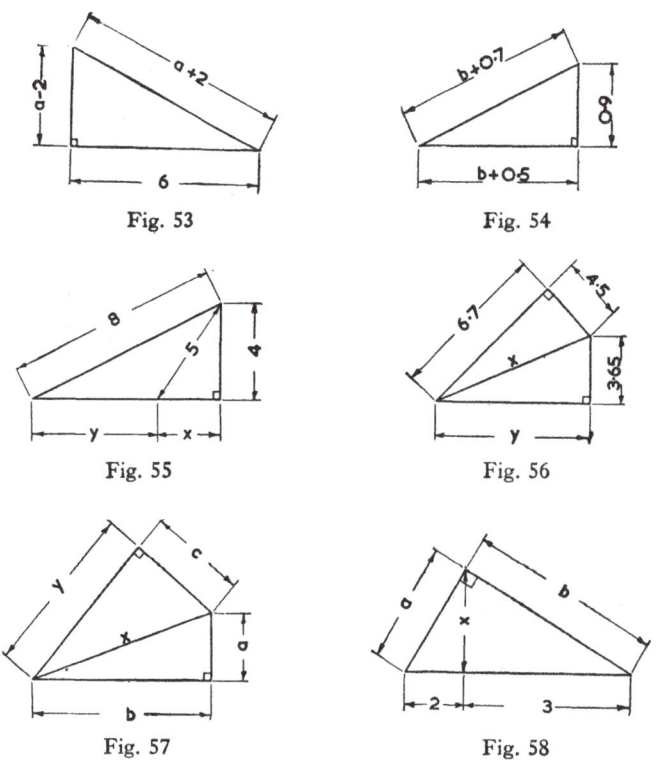

Fig. 53

Fig. 54

Fig. 55

Fig. 56

Fig. 57

Fig. 58

3. Determine the value of a in Fig. 53.

4. Determine the value of b in Fig. 54.

5. Determine the values of x and y in Fig. 55.

6. Determine the values of x and y in Fig. 56.

7. Express the values of x and y in terms of the other quantities in Fig. 57.

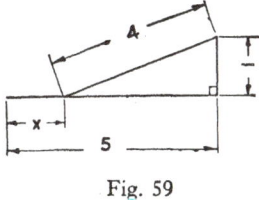

Fig. 59

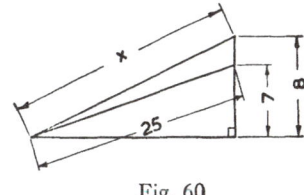

Fig. 60

8. Determine the values of a, b and x in Fig. 58.

9. Determine the value of x in Fig. 59.

10. Determine the dimension x in Fig. 60.

Exercise 31. The Theorem of Pythagoras: Practical Applications

1. Find the value of x (Fig. 61).

2. Determine the diameter D (Fig. 62).

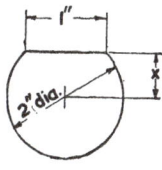

Fig. 61

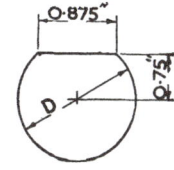

Fig. 62

3. The piston of an engine of stroke 2 ft. has a connecting rod 5 ft. long. What distance has the piston advanced along its forward stroke when the crank has moved through an angle of 90° to the direction of the piston (Fig. 63)?

4. Four holes A, B, C, D are bored in a plate. Calculate the distance between the centre of A and D if D is midway between B and C (Fig. 64).

5. A 2 in. diameter bar is rounded off with a $3\frac{1}{2}$ in. radius. Calculate the dimension x (Fig. 65).

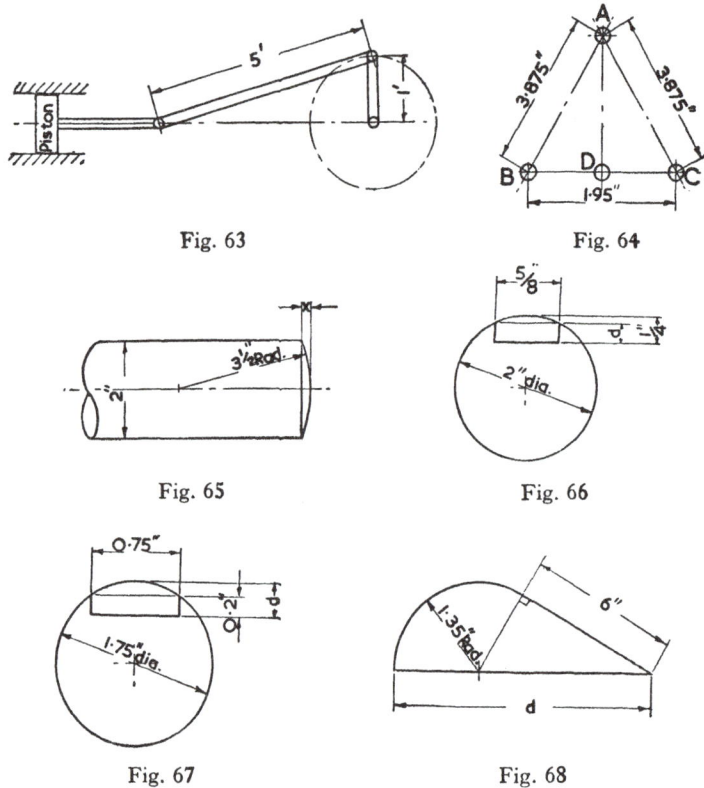

Fig. 63

Fig. 64

Fig. 65

Fig. 66

Fig. 67

Fig. 68

6. Calculate the depth d of the keyway (Fig. 66).

7. Determine the dimension d for the keyway (Fig. 67).

8. Determine the distance d (Fig. 68).

9. Determine the distance y (Fig. 69).

10. Calculate the dimension L (Fig. 70).

11. Determine the dimensions x and y (Fig. 71).

12. Holes are bored in a plate at A, B, C, D. Calculate the distances d and h (Fig. 72).

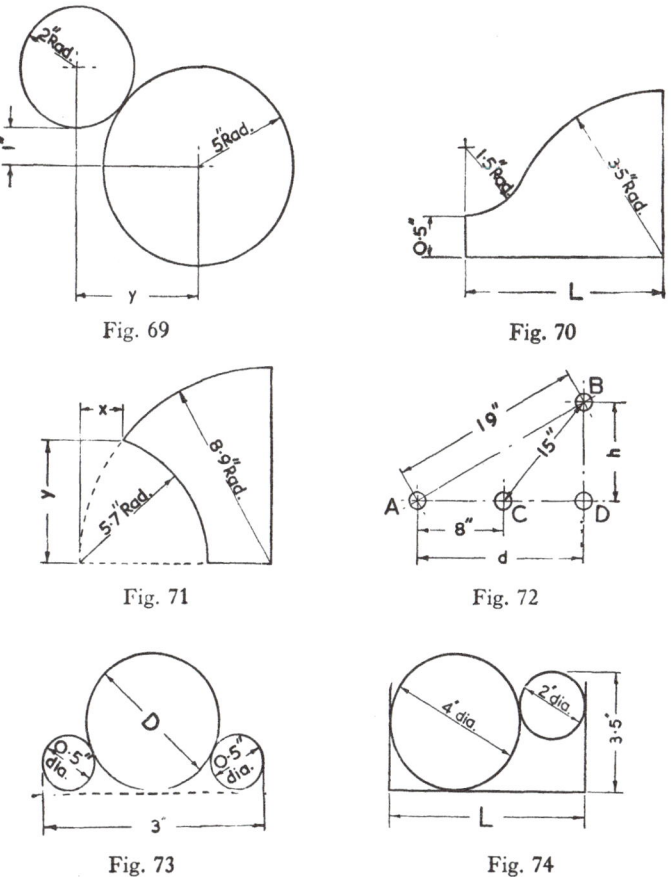

Fig. 69

Fig. 70

Fig. 71

Fig. 72

Fig. 73

Fig. 74

13. Determine the diameter D of the centre bearing (Fig. 73).

14. Determine the length of an edge of the largest metal cube which can be made from a round bar of diameter 5 in.

15. A hexagon is inscribed in a circle. If the distance between the parallel sides of the hexagon is 4 in. calculate the diameter of the circle.

16. Ball gauges are placed as shown in Fig. 74. Calculate the dimension L.

Exercise 32. Logarithms: Characteristics, Use of Tables

A

Write down the characteristics of the logarithms of the following numbers:

1. 3·142.	**2.** 17·34.	**3.** 324·9.	**4.** 3679.
5. 1000000.	**6.** 0·5086.	**7.** 0·0379.	**8.** 0·00378.
9. 0·01.	**10.** 0·000001.	**11.** 0·1005.	**12.** 1·796.
13. 0·909.	**14.** 76954·7.	**15.** 1·0.	

B

Write down the logarithms of the following:

1. 3·842.	**2.** 26.	**3.** 324.	**4.** 2986.
5. 25·42.	**6.** 2·718.	**7.** 434·3.	**8.** 89·12.
9. 1·5782.	**10.** 576100.	**11.** 0·58.	**12.** 0·0762.
13. 0·0005.	**14.** 0·00242.	**15.** 0·0000072.	

C

From the tables of antilogarithms find the numbers corresponding to the following logarithms:

1. 2·5934.	**2.** 1·4709.	**3.** $\bar{1}$·8736.	**4.** 3·1739.
5. 0·0625.	**6.** $\bar{1}$·6081.	**7.** 3·3850.	**8.** $\bar{3}$·9878.
9. 4·5973.	**10.** 0·4972.	**11.** $\bar{4}$·0000.	**12.** 0·0103.
13. 0·4343.	**14.** $\bar{1}$·6378.	**15.** 6·0000.	

Exercise 33. Logarithms: Multiplication and Division (numbers > 1)

Evaluate the following, making a rough check in each case:

1. 6.328×120.4.   2. 45.63×314.2.   3. 3.142×3.142.

4. 2.536×54.38.   5. 6.92×12.46.   6. $6.34 \times 2.593 \times 21.38$.

7. 941.6×17.51.   8. 3.174×5.689.   9. 98.78×50.57.

10. $2.576 \times 41.38 \times 5.799$.   11. 34.76×34.76.

12. $(17.95)^2 \times 3.597$.   13. $1796000 \times 32.16 \times 98.17$.

14. $1.008 \times 4.071 \times 2.108$.   15. $165.3 \div 18.18$.

16. $9654 \div 77.34$.   17. $624.3 \div 72.36$.   18. $34.82 \div 24.12$.

19. $20 \div 3.672$.   20. $(253.5 \times 19.6) \div 8.784$.

21. $\dfrac{567.2}{12.92 \times 7.987}$.   22. $\dfrac{3769}{3.142 \times 3.875 \times 3.875}$.

Exercise 34. Logarithms: Manipulation of Negative Characteristics

A

Perform the following additions:

1. $\bar{1}.6$ 2. $\bar{1}.4$ 3. $\bar{1}.5$ 4. $\bar{1}.6$ 5. $\bar{1}.7$
   $\bar{1}.2$ 1.2 $\bar{1}.8$ 2.2 $\bar{2}.9$

6. $\bar{3}.7$ 7. $\bar{4}.9$ 8. $\bar{3}.62$ 9. 0.25 10. 0.68
   3.8 3.7 $\bar{5}.84$ $\bar{2}.57$ $\bar{1}.97$

B

Perform the following subtractions:

1. $\bar{1}.8$ 2. 0.6 3. 0.6 4. $\bar{1}.3$ 5. $\bar{2}.4$
   $\bar{1}.5$ $\bar{2}.5$ $\bar{3}.8$ $\bar{3}.6$ 1.6

6. $\bar{4}.7$ 7. $\bar{5}.5$ 8. 0.72 9. 0.64 10. $\bar{1}.89$
   2.9 $\bar{3}.8$ 1.95 $\bar{1}.87$ $\bar{1}.89$

C

Evaluate

1. $\overline{3}\cdot2\times3$.   2. $\overline{2}\cdot8\times3$.   3. $\overline{1}\cdot63\times2$.   4. $\overline{3}\cdot954\times5$.

5. $\overline{2}\cdot1738\times6$.   6. $\overline{2}\cdot468\div2$.   7. $\overline{3}\cdot969\div3$.   8. $\overline{1}\cdot9766\div2$.

9. $\overline{2}\cdot3796\div3$.   10. $\overline{1}\cdot8596\div3$.

Exercise 35. Logarithms: Multiplication and Division (numbers < 1)

Evaluate the following:

1. $78\cdot35\times0\cdot8755$.

2. $297\cdot4\times0\cdot002766$.

3. $0\cdot3675\times0\cdot8642$.

4. $63\cdot42\times0\cdot9347$.

5. $0\cdot336\times0\cdot07265$.

6. $0\cdot00389\times0\cdot865$.

7. $1\cdot333\times3\cdot142\times0\cdot6248$.

8. $0\cdot3825\times38\cdot25\times0\cdot03825$.

9. $2\cdot192\div0\cdot0579$.

10. $0\cdot3471\div0\cdot0824$.

11. $0\cdot3360\div0\cdot07265$.

12. $3\cdot142\div0\cdot07634$.

13. $100\div1654$.

14. $1\div15\cdot66$.

15. $\dfrac{5\cdot625}{83\cdot9\times9\cdot34}$.

16. $\dfrac{0\cdot8234\times0\cdot0327}{0\cdot1376}$.

17. $\dfrac{16\cdot8}{17\cdot6\times38\cdot4\times0\cdot513}$.

18. $\dfrac{0\cdot04192\times0\cdot2718}{0\cdot008371\times0\cdot05913}$.

19. $\dfrac{0\cdot3183\times3\cdot142}{2\cdot718\times0\cdot3678}$.

20. $\dfrac{176\cdot8\times0\cdot125}{11\cdot05\times0\cdot4}$.

Exercise 36. Logarithms: Powers and Roots

Evaluate by logarithms:

1. $(3\cdot142)^2$.

2. $(2\cdot718)^2$.

3. $(163\cdot5)^3$.

4. $\sqrt{(796\cdot9)}$.

5. $\sqrt{(5823)}$.

6. $\sqrt[3]{(96\cdot63)}$.

7. $\sqrt[5]{(865\cdot4)}$.

8. $6\cdot45\times\sqrt{(27)}$.

9. $\sqrt[3]{(635\cdot4\times62\cdot83)}$.

10. $\sqrt[3]{(87000)}$. **11.** $(0.435)^3$. **12.** $(0.02643)^3$.

13. $(0.1856)^4$. **14.** $\sqrt{(0.1964)}$. **15.** $\sqrt{(0.001635)}$.

16. $\sqrt[3]{(0.6728)}$. **17.** $\sqrt[3]{(0.06728)}$. **18.** $318\sqrt{(0.5)}$.

19. $\sqrt{\dfrac{77.6 \times 2.395}{126.4}}$. **20.** $\sqrt{\dfrac{0.0327 \times 0.00743}{66.4 \times 0.000816}}$.

21. $\sqrt{\{(3.142)^2 + (0.8754)^2\}}$. **22.** $(0.476)^2 + \sqrt[3]{(0.983)}$.

Exercise 37. Logarithms: Miscellaneous Problems

1. If $W = \frac{1}{4}\pi d^2 l\rho$, evaluate W when $\pi = 3.142$, $d = 3.875$, $l = 36$ and $\rho = 0.26$.

2. The weight of a pipe is given by $W = \frac{\pi}{4}(D-d)(D+d)\,l\rho$. Determine the weight of a pipe when $\pi = 3.142$, $D = 3.75$ in., $d = 3.55$ in. and $l = 60$ in. if $\rho = 0.28$ lb. per cu.in.

3. A formula connected with a pendulum is $T = 2\pi\sqrt{\dfrac{l}{g}}$. Calculate the value of T when $l = 3.28$ ft., $g = 32.2$ and $\pi = 3.142$.

4. The diameter of a sphere is given by $d = \sqrt[3]{\dfrac{6V}{\pi}}$. Determine the diameter when $V = 2.248$ cu.in. $(\pi = 3.142.)$

5. The current in an electrical circuit is given by $I = \dfrac{E}{R+r}$. Calculate the value of I when $E = 1.5$ volts, $R = 30$ ohms and $r = 1.7$ ohms.

6. Calculate the value of r when $r^2 = \dfrac{6.971}{3.142 \times 8.755 \times 11.6}$.

7. A formula relating to pressures in cylinders is $D = d\sqrt{\dfrac{f+p}{f-\rho}}$. Calculate the value of D when $d = 2.5$, $f = 65.5$ and $p = 20$.

8. Evaluate $(2.718)^2 + \sqrt{0.4678}$.

9. If $E = (R+r)\sqrt{\dfrac{W}{R}}$ determine the value of E when $R = 5$, $r = 1.75$ and $W = 3.65$.

10. If $L = \dfrac{Md}{aM + b}$ determine the value of L when $M = 1.207$, $a = 5.76$, $d = 1.75$ and $b = 9.71$.

11. The velocity of water flowing through a pipe is given by the formula $v = \sqrt{\dfrac{H}{dk}}$. Find the value of v when $H = 1745$, $d = 2.375$ and $k = 0.606$.

12. A problem relating to horse-power gives
$$(\text{h.p.}) = \frac{2725 \times (4.7)^3 \times 2 \times 3.142 \times 175}{83 \times 3 \times 12 \times 33000}.$$
Calculate the h.p.

Exercise 38. Proportion: Speed, Time and Revolutions

(Work to two decimal places where necessary)

1. Show (a) that a speed of 60 miles per hour is equivalent to 88 ft. per second;

(b) that to convert m.p.h. to ft. per sec. the multiplying factor is $\frac{22}{15}$;

(c) that to convert ft. per sec. to m.p.h. the multiplying factor is $\frac{15}{22}$.

2. Convert the following speeds to ft. per sec.:

(a) 120 m.p.h.; (b) 30 m.p.h.; (c) 15 m.p.h.;

(d) 45 m.p.h.; (e) 37·5 m.p.h.

3. Convert the following speeds to m.p.h.:

(a) 66 ft. per sec. (b) 440 ft. per sec. (c) 100 ft. per sec.

(d) 40 ft. per sec. (e) 100 yd. in 10 sec.

4. An express train from London to Derby, a distance of 128 miles, takes $2\frac{3}{4}$ hours for the journey. What is the average speed in m.p.h. and in ft. per sec.?

5. A train leaves Aton at 8.55 a.m. and arrives at Beeton at 1.30 p.m., covering a distance of 135 miles. What is the average speed in m.p.h.?

6. A train travelling at an average speed of 44 m.p.h. takes 3 hr. 15 min. for a journey. What is the distance travelled?

7. An observer in a train notes that the time to pass two consecutive quarter-mile posts is t sec. Show that the average speed in m.p.h. is given by av. speed $= \dfrac{900}{t}$.

8. Calculate the speeds in m.p.h. of trains which pass consecutive quarter-mile posts in (a) 30 sec., (b) 45 sec., (c) $13\frac{1}{3}$ sec., (d) 1 min. 5 sec.

9. A train running to time covers 120 miles at an average speed of 30 m.p.h. Owing to fog it is 25 min. late. What is the actual average speed?

10. A train leaves A for D, stopping for 5 min. at both B and C. The distance $AB = 5$ miles, $BC = 15$ miles and the total journey is 40 miles. The average speed between A and B is 20 m.p.h., between B and C 30 m.p.h. and between C and D 25 m.p.h. What is the total time for the journey and the average speed?

11. How many complete turns per minute will be made by the driving wheel of a locomotive when travelling at 20 m.p.h. if the diameter of the wheel is 6 ft.?

12. The diameter of the driving wheels of a turbine locomotive is 6 ft. 6 in. and that of the front bogey wheels is 3 ft. How many revolutions does each wheel make in a minute when the speed of the locomotive is 30 m.p.h.?

13. The driving wheels of a locomotive when travelling at 30 m.p.h. make 120 revolutions per minute. Calculate the diameter of the wheels.

Exercise 39. Mensuration: Rectangle and Square

1. Find the area of a floor 17 ft. 6 in. long and 12 ft. 9 in. wide.

2. A sheet of metal is 15 in. wide. What length will have an area of 10 sq.ft.?

3. What is the length of the side of a square field of area 10 acres?

S W M

4

4. The rectangular electrode of a cell is $1\frac{3}{8}$ in. wide and has an area of $4\frac{1}{8}$ sq.in. What is its length?

5. A sheet of metal 2 ft. 3 in. square weighs 4 lb. per sq.ft. What is the weight of the sheet?

6. An open tank 10 ft. long, 7 ft. wide and 7 ft. deep is lined with sheet metal costing 7s. 6d. a sq.yd. What is the cost of the lining?

7. Mild steel of rectangular section $\frac{3}{4}$ in. by $\frac{5}{8}$ in. broke under a test of 30,000 lb. What is the tensile strength in lb. per sq.in.?

8. What length of angle iron would be needed to reinforce the edges of a closed box of dimensions 16 in. by 8 in. by 6 in.?

9. What length of boarding $4\frac{1}{2}$ in. wide would be needed to floor a room 13 ft. 6 in. long and 12 ft. 6 in. wide?

10. The external dimensions of a window frame are 4 ft. 6 in. by 1 ft. 6 in. and the wood is 2 in. wide. Calculate (a) the area of one side of the frame, (b) the area of glass visible.

11. A wall of a room is 18 ft. long and 13 ft. 6 in. high. It contains two windows and a door. Each window is 2 ft. 3 in. wide and 5 ft. high. The door is 2 ft. 6 in. wide and 6 ft. 6 in. high. Calculate the area of the wall.

12. A rectangular metal plate of dimensions $10\frac{1}{8}$ in. by $8\frac{3}{8}$ in. is melted down and recast into a square plate of equal area. What is the perimeter of the square correct to two places of decimals?

Exercise 40. Mensuration: Areas of Sections, Weights per Foot Run

Calculate (a) the area of each of the sections shown in Figs. 75–84; (b) the weight per foot run of girders having these sections, with respective densities given in parentheses.

1. Fig. 75. (0·28 lb. per cu.in.) **2.** Fig. 76. (0·28 lb. per cu.in.)

3. Fig. 77. (0·26 lb. per cu.in.) **4.** Fig. 78. ($\frac{1}{4}$ lb. per cu.in.)

5. Fig. 79. (0·24 lb. per cu.in.) **6.** Fig. 80. (0·28 lb. per cu.in.)

7. Fig. 81. (0·28 lb. per cu.in.) **8.** Fig. 82. (0·26 lb. per cu.in.)

9. Fig. 83. (0·28 lb. per cu.in.) **10.** Fig. 84. (0·25 lb. per cu.in.)

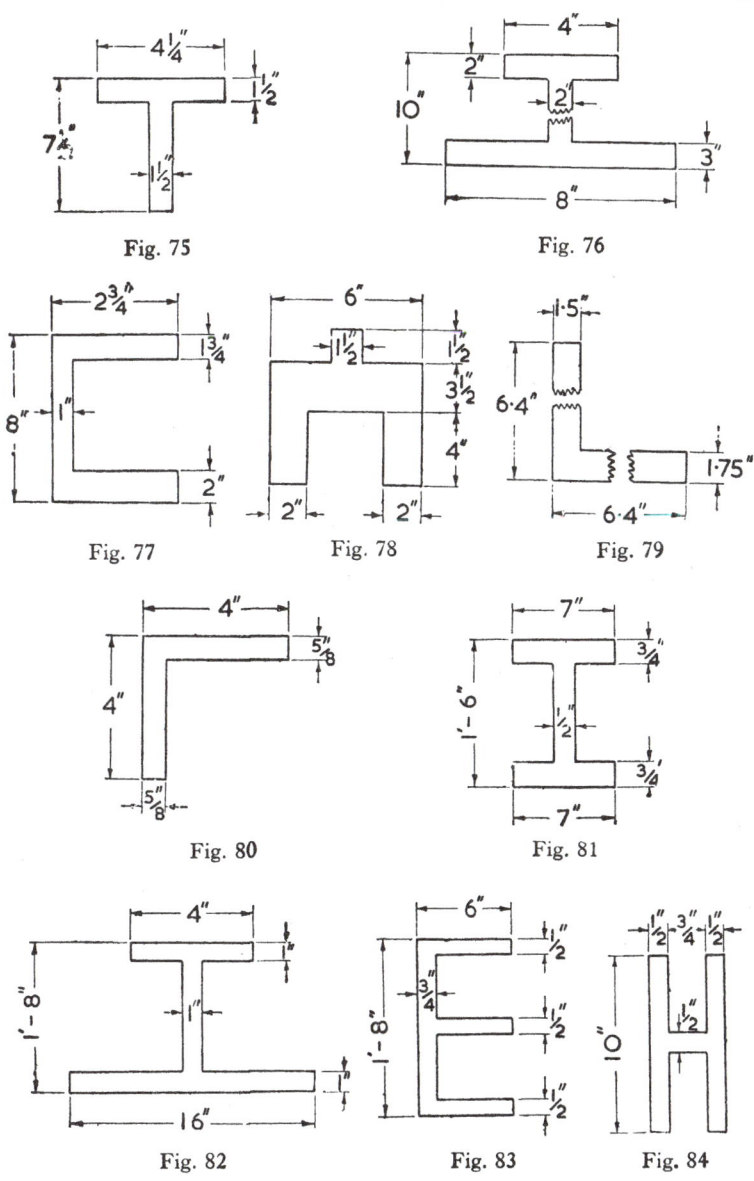

Fig. 75

Fig. 76

Fig. 77

Fig. 78

Fig. 79

Fig. 80

Fig. 81

Fig. 82

Fig. 83

Fig. 84

Exercise 41. Mensuration: Rectangular Solid

(*Logarithms should be used when necessary*)

1. A rectangular tank of dimensions 10·75 in. by 8·5 in. by 7·25 in. is filled with water. Calculate its capacity (*a*) in cu.in., (*b*) in gallons. (*Note.* 1 gallon occupies 277·3 cu.in.)

2. A bar of steel is 3 in. square. If the steel weighs 0·28 lb.cu.in., find what length of bar will weigh 1 cwt.

3. Find the total area of steel plates required to make a closed rectangular tank of dimensions 12 ft. by 9 ft. by 7 ft. How many gallons of water would the tank hold? (1 gallon occupies 277·3 cu.in.)

4. Air passes at the rate of 660 cu.ft. per min. into a room of dimensions 20 ft. by 15 ft. by 12 ft. How many times is the air changed per hour?

5. What is the depth of water in a tank 10 ft. long and 6 ft. wide if it holds 240 cu.ft. of water?

6. A bar of copper is 5 in. wide and $1\frac{1}{4}$ in. thick and weighs 72 lb. If 1 cu.in. of copper weighs 0·32 lb., what is the length of the bar?

7. The external dimensions of an open stone trough are 36 in. by 20 in. by 15 in. deep. The sides are 2 in. thick and the bottom 3 in. thick. Find the weight of the trough if 1 cu.ft. of stone weighs 180 lb.

8. How many sq.ft. of steel plates are needed to construct a tank of dimensions 10 ft. by 8 ft. by 6 ft. allowing $\frac{1}{10}$ of the whole for overlap?

9. The area of the surface of a cube is 3456 sq.in. What is its volume in cu.ft.?

10. A metal box with lid, 14 in. wide and 12 in. deep, is to be made from 1168 sq.in. of sheet metal. What will be the length of the box?

Exercise 42. Mensuration: Triangle and Triangular Prism

Note. Area of triangle$=\frac{1}{2}bh$ or $\sqrt{\{s(s-a)(s-b)(s-c)\}}$.

1. Find the area of a triangle whose base is 11·5 in. and whose altitude is 8 in.

2. Determine the area and weight of a triangular metal plate whose base is 24 in. and altitude 18 in. if the metal weighs 4 lb. per sq.ft.

3. A right-angled triangular metal plate has the hypotenuse 10 in. long and one of the other sides 7·85 in. long. The metal weighs 5 lb. per sq.ft. Calculate the area and weight of the plate.

4. The area of a triangle is 8·1 sq.in. and its altitude 4·5 in. What is the length of the base?

5. Determine the altitude of a triangle whose base is 6 ft. long and whose area is $9\frac{3}{4}$ sq.ft.

6. Find the area of a triangle whose sides are 5 in., 8 in. and 9 in.

7. Find the weight of a triangular metal plate whose sides are 6·5 in., 7·76 in. and 8·5 in. if the metal weighs 0·028 lb. per sq.in.

8. A prism 12 in. long has a section in the form of a triangle whose sides are 5 in., 5 in. and 8 in. Calculate the volume of the prism.

9. Determine the weight of a metal wedge 10 in. long, whose section is in the form of a triangle of sides 5 in., 6 in. and 7 in., the density of the metal being 0·28 lb.cu.in.

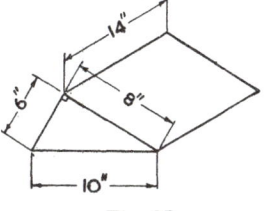

Fig. 85

10. A rectangular block of metal of dimensions 10 in. by 8 in. by 8 in. is melted down and recast into a triangular prism whose section is an equilateral triangle of sides 6 in. Neglecting waste find the length of the prism made.

11. A wedge (Fig. 85) and a cube have the same volume. Calculate the length of the side of the cube.

12. A wedge 12 in. long has a section in the form of a triangle whose sides are 6·4 in., 6·1 in. and 3·5 in. Find its weight if 1 cu.in. of metal weighs 0·29 lb.

Exercise 43. Mensuration: Trapezium

1. A trapezium whose parallel sides are 55 in. and 35 in. has an altitude of 20 in. What is its area?

2. Fig. 86 represents a pane of glass. How many sq.ft. are in four such panes?

3. A channel section is 6 ft. wide at the top and 4 ft. 4 in. wide at the bottom. If it is 3 ft. deep, how many cu.ft. of earth must be removed to dig a channel 10 ft. long?

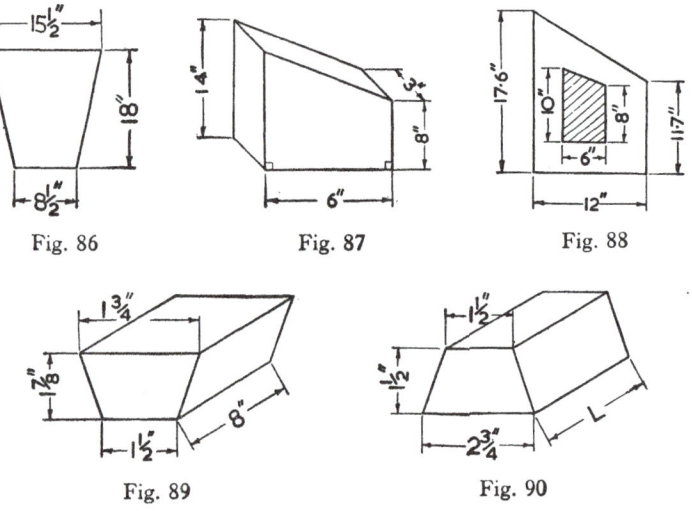

Fig. 86   Fig. 87   Fig. 88

Fig. 89   Fig. 90

4. Calculate the weight of the block (Fig. 87) if the density of the material is 0·28 lb.cu.in.

5. Fig. 88 represents the section of a block which is 2 in. thick and weighs 72 lb. after a portion (shown shaded) is removed. Determine the density of the material in lb.cu.in.

6. Determine the weight of the ingot (Fig. 89) if the density of the metal is 0·28 lb.cu.in.

7. Determine the length L of the ingot (Fig. 90) if it weighs 25 lb. and the density of the material is 0·26 lb.cu.in.

8. An ingot 14 in. long has a section in the form of a trapezium 2 in. wide at the top, $1\frac{1}{4}$ in. wide at the bottom and $2\frac{1}{8}$ in. high. Calculate the density of material if it weighs 19·36 lb.

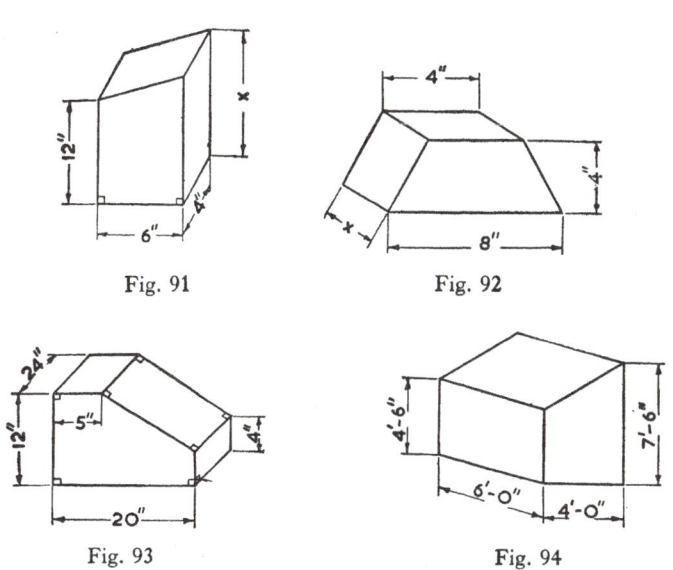

Fig. 91   Fig. 92

Fig. 93   Fig. 94

9. The block (Fig. 91) is composed of an alloy of density $\frac{1}{4}$ lb.cu.in. If it weighs 75 lb. calculate the dimension x.

10. A steel anchor block (Fig. 92) of density 480 lb.cu.ft. weighs 40 lb. Calculate its thickness x in inches.

11. Fig. 93 represents the top of a desk. Calculate (*a*) the area of the end of the desk, (*b*) the volume in cu.ft., (*c*) the area of the sloping top in sq.ft.

12. Fig. 94 represents a shed. Calculate (*a*) the volume in cu.ft., (*b*) the total surface area of the top, front, sides and back in sq.ft.

13. Calculate the area of the template (Fig. 95).

14. Determine the area of the metal plate (Fig. 96).

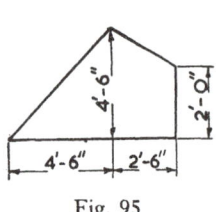

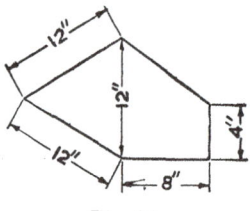

Fig. 95   Fig. 96

Mensuration: Circle and Annulus

Special attention should be given to the following before attempting the exercise.

Circumference of circle $= 2\pi r = \pi d$.

$$\text{Area of circle} = \pi r^2. \tag{1}$$

$$= \frac{\pi}{4} d^2. \tag{2}$$

$$\text{Area of annulus} = \pi(R^2 - r^2). \tag{3}$$

$$= \pi(R - r)(R + r). \tag{4}$$

$$= \frac{\pi}{4}(D^2 - d^2). \tag{5}$$

$$= \frac{\pi}{4}(D - d)(D + d). \tag{6}$$

Note. (*a*) Equations (4) or (6) should be used when finding the area of an annulus.

(*b*) Equations (3) or (5) should be used when finding the values of the radii R, r, or the diameters D, d.

Exercise 44. Mensuration: Circle and Annulus

1. Determine the circumference of a circle when (*a*) $r = 1\cdot5$ in., (*b*) $r = 2\cdot25$ in., (*c*) $d = 3\cdot75$ in., (*d*) $d = 1\frac{3}{8}$ ft.

2. The circumference of a circle is $18\cdot75$ in. What is its diameter?

3. How many times will the driving wheel of a locomotive turn in travelling 1 mile, if the wheel diameter is 5 ft. 6 in.?

4. A flywheel of diameter 3 ft. 6 in. makes 240 r.p.m. What is the speed of a point on the rim in ft. per sec.?

5. A point on the rim of a wheel making 1800 r.p.m. has a speed of 60 ft. per sec. What is the diameter of the wheel in feet?

6. Determine the area of circles when (a) $r = 4.75$ in., (b) $r = 3\frac{7}{8}$ in., (c) $d = 5.6$ ft.

7. A round bar has a cross-section of 175 sq.in. What is the diameter of the bar?

8. A metal disk of 6 in. diameter is melted down and recast into disks of 3 in. diameter and of the same thickness as the original disk. How many small disks can be made?

9. The pressure of steam on a piston of 18 in. diameter is 80 lb. per sq.in. What is the total pressure on the piston in tons?

10. Calculate the weight of a metal plate of 6 in. diameter bored with four holes of $\frac{3}{4}$ in. diameter if the metal weighs 0·35 lb. per sq.in. of surface.

11. Determine the area of the face of a large washer of external diameter 3 in. and internal diameter 2 in.

12. A cast-iron pipe has a bore of 1·75 in. and is 1 in. thick. Calculate the area of the cross-section.

13. The area of one face of a washer is $9\frac{1}{2}$ sq.in. and the internal diameter is 2 in. Calculate the breadth of the washer to the nearest whole number.

14. The cross-section of a pipe is 5·645 sq.in. and the internal diameter is $2\frac{1}{4}$ in. What is the thickness of the metal?

Mensuration: Cylinders and Pipes

Note. Curved surface area of cylinder $= 2\pi r l = \pi d l$.

Total surface area of cylinder $\quad = 2\pi r l + 2\pi r^2$

$$= \pi d l + \pi \frac{d^2}{2}.$$

Exercise 45. Mensuration: Heating Surface of Tubes: Volume and Weight of Bars and Tubes

1. Calculate the volume and curved surface area of a round bar of diameter 5 in. and length 12 in.

2. For the following locomotives calculate (i) the heating surface of the large and small smoke tubes, (ii) the total heating surface.

(a) 6P Class. Royal Scot. (Taper boiler.)
   Large tubes: 28, $5\frac{1}{8}$ in. diameter. Length 14 ft. 3 in.
   Small tubes: 180, $1\frac{7}{8}$ in. diameter. Length 14 ft. 3 in.

(b) A Class. Super Pacific.
   Large tubes: 32, $5\frac{1}{4}$ in. diameter. Length 19 ft.
   Small tubes: 168, $2\frac{1}{4}$ in. diameter. Length 19 ft.

(c) 5XP Class. Patriot.
   Large tubes: 24, $5\frac{1}{4}$ in. diameter. Length 14 ft.
   Small tubes: 140, $2\frac{1}{8}$ in. diameter. Length 14 ft.

(d) 2P Class. (Simple.)
   Large tubes: 21, $5\frac{1}{8}$ in. diameter. Length 10 ft. $10\frac{1}{2}$ in.
   Small tubes: 146, $1\frac{3}{4}$ in. diameter. Length 10 ft. $10\frac{1}{2}$ in.

3. Determine the separate and combined heating surfaces of the tubes for the locomotive:

Large tubes: 18, $5\frac{1}{8}$ in. diameter. Length 12 ft. 3 in.
Small tubes: 148, $1\frac{3}{4}$ in. diameter. Length 12 ft. 3 in.
Superheater, triple elements: 18, $1\frac{1}{8}$ in. diameter. Length 11 ft.

4. Determine the weight of a round bar 10 in. long and of diameter 2·5 in. if the material weighs 0·26 lb.cu.in.

5. A portion of shafting 8 in. long has a volume of 100 cu.in. Calculate the diameter of the shaft.

6. A steel rod of diameter 3 in. weighs 100 lb. If 1 cu.in. of steel weighs 0·28 lb. calculate the length of the rod.

7. Determine (a) the length of a bar of section 4 sq.in. and having a volume of 1 cu.ft., (b) the length of a round bar of section 2 sq.in. and having a volume of 1 cu.ft.

8. A cu.ft. of metal is made into a bar of section 0·16 sq.in., while a second cu.ft. of similar metal is made into a bar of radius 0·2 in. What is the difference in length of the bars?

9. A solid steel bar of 1½ in. diameter is to be replaced by a steel tube of the same weight and length and whose external diameter is 3 in. Calculate the thickness of the tube.

10. A cast-iron pipe weighs 1 cwt. If the external diameter is 2·75 in. and internal diameter 2·25 in., what is the length of the pipe if 1 cu.in. of iron weighs 0·26 lb.?

11. Calculate the length of a steel tube of bore 3·2 in. and thickness 0·3 in. if it weighs 1 cwt. and the density of steel is 0·28 lb. per cu.in.

Exercise 46. Mensuration: Capacity and Rate of Flow of Water in Pipes

1. If 1 cu.ft. of water weighs 62·3 lb. and a gallon of water weighs 10 lb., show that 1 gallon is equivalent to 0·1605 cu.ft. or 277·3 cu.in.

2. A water tank is 5 ft. long, 4 ft. wide and holds water to a depth of 3 ft. How many gallons are in the tank?

3. A pipe from a water tower discharges 3500 gallons of water in 5 min. when the rate of flow is 5 ft. per sec. What is the diameter of the pipe in inches?

4. Water flows through a pipe of 3½ in. diameter and length 3 ft. in 1 sec. How many gallons pass through the pipe in a minute?

5. If 32,710 gallons of water are pumped into a tank of length 35 ft. and width 15 ft., to what height will the water rise?

6. A water tower of diameter 10 ft. is filled to a depth of 7 ft. How many gallons does it contain?

7. If 3000 gallons of water are pumped into a tender in 5 min. through a pipe of diameter 7 in., what is the rate of flow in ft. per sec.?

8. A locomotive has a water capacity of 4000 gallons. How long will it take to water it through a pipe of diameter 7 in. if the rate of flow is 6 ft. per sec.?

9. What is the capacity in gallons of a boiler of length 20 ft. 3 in. and diameter 5 ft. 8 in.? (Internal dimensions.)

10. Water flows at 10 ft. per sec. through a channel 12 in. wide at the top, 8 in. wide at the bottom and 6 in. deep. What is the rate of flow in gallons per minute?

Exercise 47. Mensuration: Miscellaneous Problems

1. Determine the weight of a steel tyre of external diameter 5 ft., internal diameter 4 ft. 6 in. and 2 in. thick if the density of steel is 470 lb.cu.ft.

2. An ingot whose section is in the form of a trapezium 8 in. wide at the top, 12 in. wide at the bottom, 4 in. high and 5 in. thick is melted down and recast into a bar of diameter $3\frac{1}{2}$ in. What length of bar can be made?

3. A round bar 10 in. long and of diameter 2 in. is melted and recast into a pipe of the same length and of external diameter 3 in. What is the bore of the pipe?

4. Calculate the weight per foot of a metal tube of bore 2·3 in. and 0·2 in. thick if 1 cu.in. of metal weighs 0·26 lb.

5. What is the length of a steel pipe of external diameter $3\frac{1}{2}$ in. and internal diameter $2\frac{1}{2}$ in. if the density of steel is 0·28 lb.cu.in., and the pipe weighs 13·2 lb.?

6. A flywheel of $3\frac{1}{2}$ ft. diameter makes 2000 r.p.m. What is the speed of a point on the rim in ft. per sec.?

7. A solid wheel of diameter $1\frac{1}{4}$ ft. makes 500 r.p.m. What is the speed of the rim in ft. per sec.? If it weighs 800 lb. and the density of the metal is 100 lb.cu.ft. calculate its thickness.

8. Determine the weight of the channel section (Fig. 97) if 1 cu.in. of metal weighs 0·34 lb., giving your answer to the nearest ton.

9. How many castings (Fig. 98) can be made from 1 ton of metal of density 0·28 lb.cu.in.?

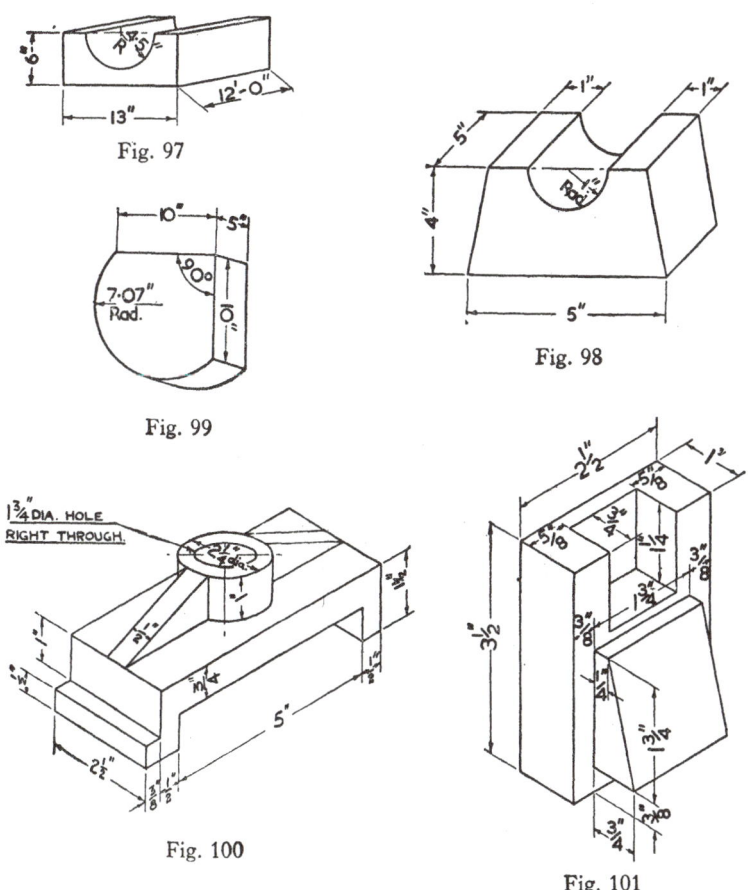

Fig. 97

Fig. 98

Fig. 99

Fig. 100

Fig. 101

10. Determine the volume of the casting (Fig. 99).

11. Determine the total volume of the casting (Fig. 100).

12. Determine the volume of the solid shown in Fig. 101.

Mensuration: Cone, Frustum and Sphere

Note. Sphere of radius r, diameter d.

(a) Surface area $= 4\pi r^2 = \pi d^2$.

(b) Volume $= \frac{4}{3}\pi r^3 = \frac{1}{6}\pi d^3$.

Cone of radius r, height h, slant height l.

(a) Volume $= \frac{1}{3}\pi r^2 h$.

(b) Curved surface area $= \pi r l$.

(c) Total surface area $= \pi r l + \pi r^2$.

Frustum of cone. Radius of top $= r$, radius of bottom $= R$, height $= h$, slant height $= L$.

(a) Volume of frustum $= \dfrac{\pi h}{3}\,(R^2 + Rr + r^2)$.

(b) Area of curved surface $= \pi L(R+r)$.

Exercise 48. Mensuration: Cone, Frustum and Sphere

1. Determine the volume and the curved surface area of a cone when

(a) $r = 7$ in., $h = 24$ in.;   (b) $r = 7{\cdot}5$ in., $h = 5{\cdot}7$ in.;

(c) $r = 5$ in., $l = 13$ in.;   (d) $h = 3$ in., $l = 7$ in.

2. Determine the volume and surface area of a sphere when

(a) $r = 3$ in.;   (b) $r = 6{\cdot}5$ in.;   (c) $r = 13$ in.

3. Calculate the total weight of 1000 ball bearings of diameter $\frac{3}{4}$ in. if 1 cu.in. of metal weighs 0·28 lb.

4. How many steel balls of diameter 0·375 in. can be made from 1 cwt. of steel if the density is 0·28 lb.cu.in.?

5. A metal ball, used in conjunction with a magnetic crane for breaking down metal, weighs 1 ton. If the density of the metal ball is 500 lb.cu.ft., calculate the diameter of the ball.

6. A navigation buoy consists of a cone surmounting a hemisphere. The total height is 12 ft. 9 in. and the greatest diameter is 4 ft. 6 in. Determine the volume of the buoy.

7. A cone of height 16 in. and base radius 4 in. is melted down and recast into a solid ball. Determine the diameter of the ball.

8. A cylindrical boiler has hemispherical ends. The total length is 15 ft. and the diameter 6 ft. What is the capacity and how many sq.ft. of metal would be needed to make the boiler?

9. A hollow spherical shell of internal diameter 4 in. weighs 28 lb. Calculate the thickness of the metal if 1 cu.in. weighs 0·26 lb.

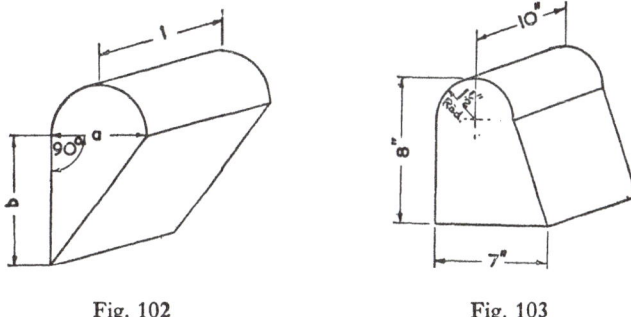

Fig. 102   Fig. 103

10. Write down the formula for the area of the cross-section of the casting (Fig. 102) in terms of a, b and l. Calculate the volume of the casting if $a = 4$ ft., $b = 5$ ft. and $l = 10$ ft.

11. Calculate the weight of the casting (Fig. 103) if 1 cu.in. of the metal weighs 0·29 lb.

12. Two spheres of radius 3 in. and 2 in. respectively are to be replaced by a single sphere having the same volume as the two together. Calculate the radius of the single sphere.

13. A bucket 15 in. high has a rim diameter of 12 in. and base diameter 10 in. How many gallons of water will it hold? (1 gallon occupies 277·3 cu.in.)

14. A casting in the form of a frustum of a cone is 2 ft. long. Its end diameters are 3 ft. and 1 ft. respectively. What is its weight if 1 cu.ft. of the metal weighs 300 lb.?

15. How many gauges (Fig. 104) can be made from 1 cu.ft. of metal?

16. Determine the weight of the solid (Fig. 105) if the density of the material is 450 lb.cu.ft.

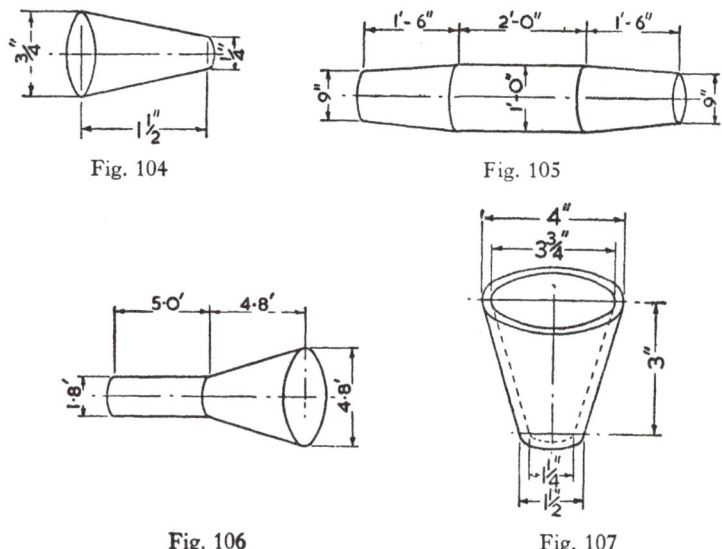

Fig. 104

Fig. 105

Fig. 106

Fig. 107

17. The casting (Fig. 106) weighs 1·403 tons. Determine the density of the material in lb. per cu.ft.

18. Determine the weight of the glass oil filter (Fig. 107) if 1 cu.in. of glass weighs 0·09 lb.

19. A solid metal cone of diameter 10 in. and height 12 in. is melted and recast into a solid in the form of a frustum of a cone of base diameter 10 in. and top diameter 6 in. What is the altitude of the frustum?

20. A churn in the form of a frustum of a cone has a rim diameter of 8 in. and base diameter 20 in. If it holds 15 gallons calculate the height of the churn if 1 gallon occupies 0·156 cu.ft.

Mensuration: Arcs, Sectors and Segments

Note.

(a) Length of minor arc $= \dfrac{\theta}{360}$ of circumference.

$$= \frac{\theta}{360} \times 2\pi r. \quad \text{(Sketch } (a).)$$

(b) Length of major arc $= \dfrac{\theta}{360} \times 2\pi r. \quad$ (Sketch (b).)

(c) Area of minor sector $= \dfrac{\theta}{360}$ of area of circle.

$$= \frac{\theta}{360} \times \pi r^2. \quad \text{(Sketch } (c).)$$

(d) Area of major sector $= \dfrac{\theta}{360} \times \pi r^2. \quad$ (Sketch (d).)

(e) Area of minor segment = area of minor sector − area of the triangle. (Sketch (e).)

(f) Area of major segment = area of major sector + area of the triangle. (Sketch (f).)

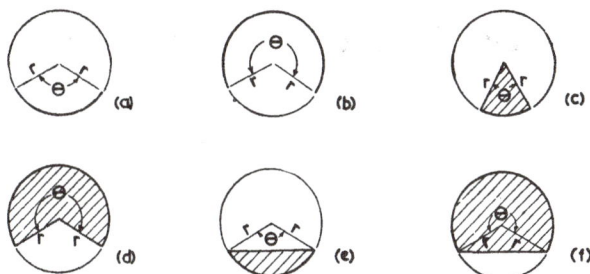

Exercise 49. Mensuration: Arcs, Sectors, Segments

1. An arc subtends an angle of 60° at the centre of a circle of radius 5 in. Calculate the length of the arc and the area of the minor sector.

2. An arc subtends an angle of 45° at the centre of a circle of radius 6 in. Calculate the length of the arc and the area of the sector.

3. At the centre of a circle of radius 3 in. an arc subtends an angle of 120°. Determine the length of the major arc and the area of the major sector.

4. An arc 6 in. long is cut off by two radii in a circle of radius 4 in. Calculate the angle at the centre (to the nearest degree) subtended by this arc. What is the area of the sector formed?

5. Determine the area of the shaded portion in Fig. 108.

6. The area of the sector of a circle of radius 4 in. is 10·47 sq.in. Determine the angle of the sector.

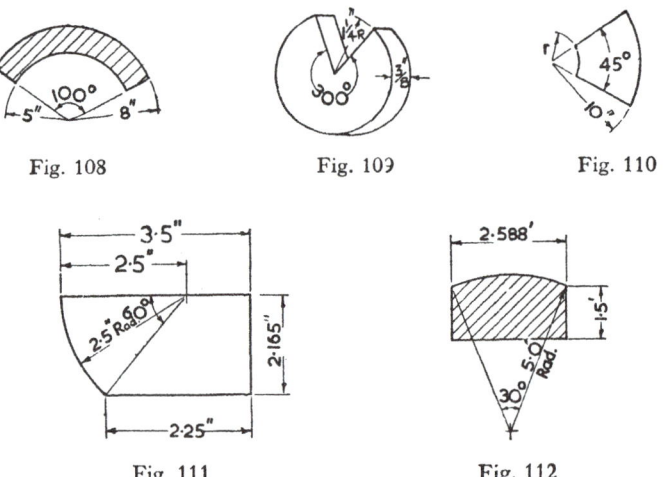

Fig. 108   Fig. 109   Fig. 110

Fig. 111   Fig. 112

7. Determine the radius of an arc which subtends an angle of 210° at the centre of a circle if the area of the sector formed is 66 sq.in.

8. Calculate the weight of the rocker (Fig. 109) if the density of the metal is 0·28 lb.cu.in.

9. The vane for a fan is to be made from a sector of metal of radius 10 in. The area of the vane is to be 38 sq.in. Calculate the radius r of the metal to be cut off (Fig. 110).

10. Determine the area of the template (Fig. 111).

11. A chord subtends an angle of 108° at the centre of a circle of radius 5 in. Calculate
  (*a*) the length of the minor arc;
  (*b*) the area of the minor sector;
  (*c*) the area of the triangle formed by the chord and radii;
  (*d*) the area of the minor segment.

12. A 1 in. flat is machined on a bar of diameter 2 in. and length 5 in. What weight of metal is removed if the density of the metal is 0·25 lb.cu.in.?

13. Determine the area of the section shown in Fig. 112.

14. Determine the angle subtended at the centre of a circle of radius 1·414 in. by a chord of length 2 in.

15. A metal bar of hexagonal section is to be 12 in. long and made from a round bar of diameter 8 in. Calculate the volume of metal to be removed.

Trigonometry

Note.

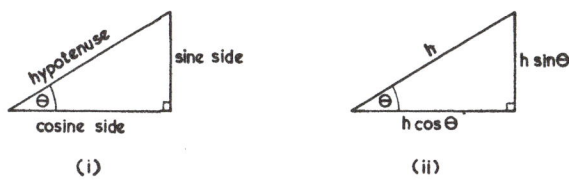

(i)   (ii)

sine side $= h \sin \theta.$ cosine side $= h \cos \theta.$

$$\frac{\text{sine side}}{\text{cosine side}} = \frac{\sin \theta}{\cos \theta} = \tan \theta.$$

Exercise 50. Trigonometry: Sine, Cosine and Tangent

1. In the following right-angled triangles express the lengths of the sine side and the cosine side in terms of the hypotenuse and the given angle.

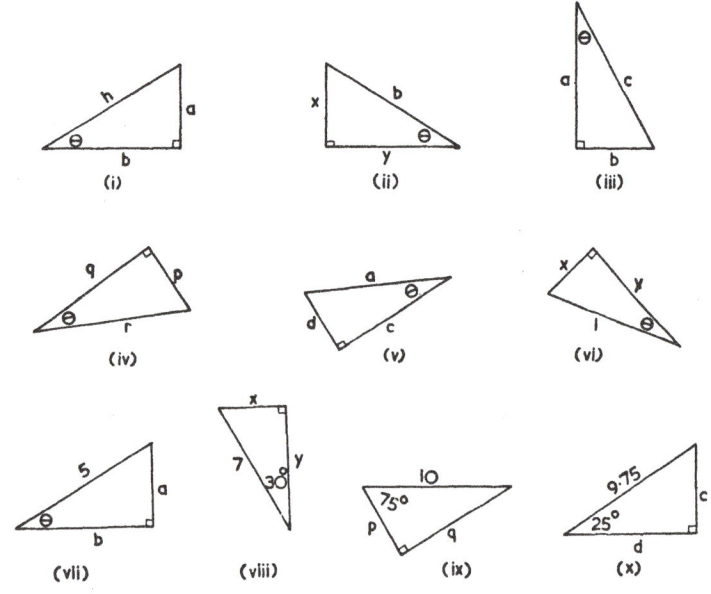

2. Write down the ratios for sin θ, cos θ and tan θ for the following triangles:

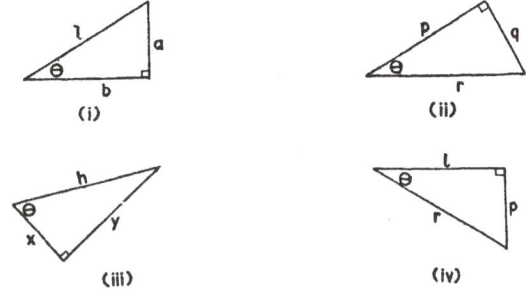

3. Write down the ratios for sin θ, cos θ and tan θ for the following triangles:

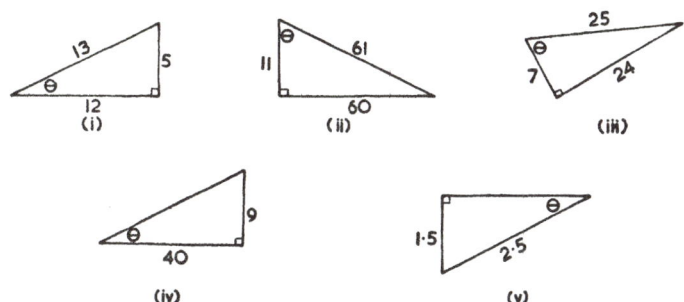

4. A 30°, 60° set-square has its longest edge 10 in. Determine the lengths of the other edges, given sin 30° = 0·5 and cos 30° = 0·866.

5. If sin 60° = 0·866 and cos 60° = 0·5, determine the lengths of the shorter sides of a triangle whose angles are 30°, 60° and 90° if the greatest side is 2 in. long.

6. Given sin 45° = cos 45° = 0·7071, determine the lengths of the shorter sides of a 45° set-square whose longest edge is 10 in.

7. In a right-angled triangle BAC, the short side $AB = 3$ in. and tan $B = 1·7321$. Determine the length of the side AC.

8. Given sin $\theta = \frac{1}{3}$, determine the values of cos θ and tan θ.

9. If cos $\theta = \frac{12}{13}$, determine the values of sin θ and tan θ, giving the answers as fractions.

10. Given tan $\theta = 0·3$, evaluate sin θ and cos θ.

Exercise 51. Trigonometry: Use of Tables

1. Using your tables, tabulate the values of sin θ, cos θ and tan θ when θ has the following values:

15°, 20°, 30°, 45°, 50°, 70°, 75°, 33°, 47° 62°.

2. Tabulate the values of $\sin\theta$, $\cos\theta$ and $\tan\theta$, when θ has the following values:

$$0°\ 18', \quad 5°\ 27', \quad 83°\ 38', \quad 47°\ 45', \quad 89°\ 24',$$
$$0°\ 59', \quad 37°\ 15', \quad 69°\ 59', \quad 3°\ 19', \quad 36°\ 52'.$$

3. If (a) $\sin\theta=0\cdot6$, calculate θ. (b) $\sin\theta=0\cdot2359$, calculate θ.
(c) $\sin\theta=0\cdot75$, calculate θ. (d) $\sin\theta=0\cdot9359$, calculate θ.
(e) $\sin\theta=0\cdot0166$, calculate θ.

4. Determine the angle θ when
(a) $\cos\theta=0\cdot8$, (b) $\cos\theta=0\cdot9179$,
(c) $\cos\theta=0\cdot6989$, (d) $\cos\theta=0\cdot2843$,
(e) $\cos\theta=0\cdot002$.

5. Determine the angle θ when
(a) $\tan\theta=1\cdot0000$, (b) $\tan\theta=0\cdot011$,
(c) $\tan\theta=0\cdot6856$, (d) $\tan\theta=2\cdot6984$,
(e) $\tan\theta=6\cdot7725$.

6. Tabulate the log sin, log cos and log tan of the following angles:
$$5°\ 30', \quad 30°, \quad 47°\ 27', \quad 62°\ 19', \quad 85°\ 53', \quad 88°\ 30', \quad 89°\ 12'.$$

7. Calculate the angle θ when
(a) $\log\sin\theta=\bar{1}\cdot4323$, (b) $\log\cos\theta=\bar{1}\cdot6174$,
(c) $\log\tan\theta=0\cdot0459$, (d) $\log\cos\theta=\bar{1}\cdot9864$,
(e) $\log\sin\theta=\bar{1}\cdot9745$.

8. Use your tables to verify the following:
$\sin 30°=\cos 60°$, $\sin 43°\ 45'=\cos 46°\ 15'$, $\sin 73°\ 42'=\cos 16°\ 18'$.

Exercise 52. Trigonometry: Simple Problems

In the triangle (Fig. 113):

1. If $c=5$ in., $A=40°$, determine the value of a, b and B.

2. If $c=11\cdot7$ in., $A=36°$, determine the value of a, b and B.

3. If $c=8\cdot75$ in., $A=32°\ 18'$, determine the value of a, b and B.

4. If $b = 8$ in., $A = 48°$, determine the value of a, c and B.

5. If $a = 6·3$ in., $A = 35°$, determine the value of b, c and B.

6. If $c = 7·9$ in., $a = 3·5$ in., determine the value of A.

7. If $c = 8·96$ in., $b = 5·63$ in., determine the value of A.

8. If $c = 15$ in., $a = 12·65$ in., determine the value of B.

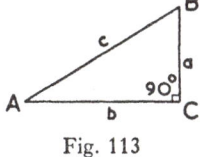

Fig. 113

9. If $a = 4·75$ in., $b = 6·17$ in., determine the value of c and A.

10. If $a = 15·68$ in., $b = 24·75$ in., determine the value of B.

Exercise 53. Trigonometry: Workshop Problems

1. Determine the angle A of the template (Fig. 114).

2. Determine the angle A and the dimension x (Fig. 115).

3. Calculate the lengths x and y (Fig. 116).

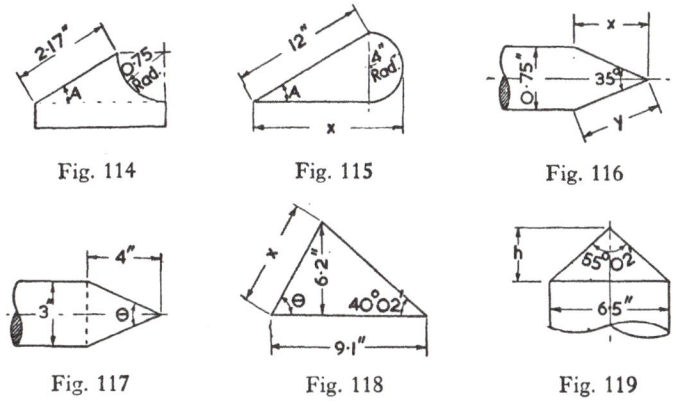

Fig. 114   Fig. 115   Fig. 116

Fig. 117   Fig. 118   Fig. 119

4. Calculate the angle θ (Fig. 117).

5. Determine the angle θ and the length x (Fig. 118).

6. Determine the dimension h (Fig. 119).

7. Calculate the diameter *d* (Fig. 120).

8. Calculate the height *h* (Fig. 121).

9. Determine the angle θ (Fig. 122).

10. Three holes of diameter 1 in. are bored in a metal plate. Determine the angle θ and the dimension *x* (Fig. 123).

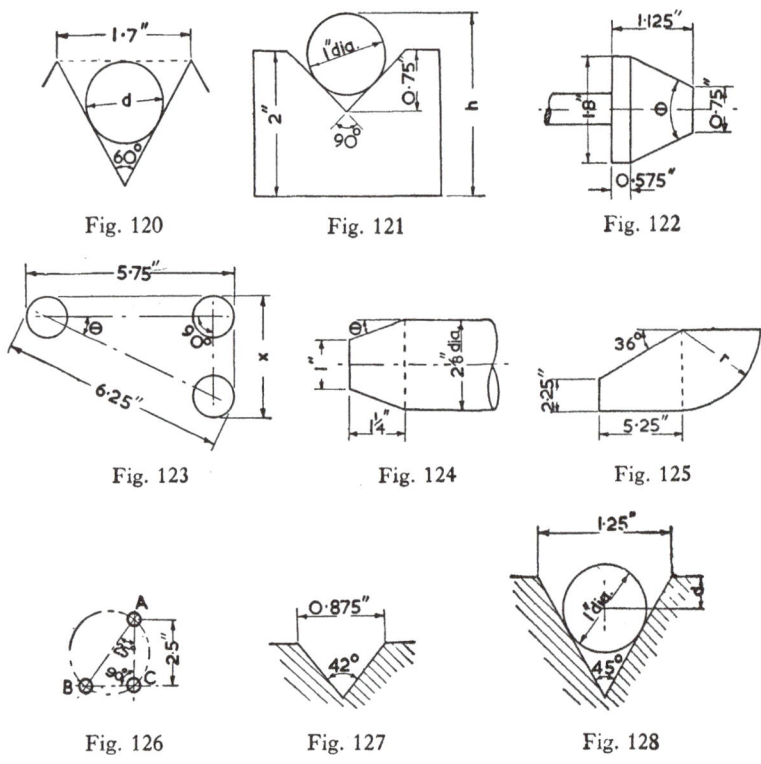

Fig. 120   Fig. 121   Fig. 122

Fig. 123   Fig. 124   Fig. 125

Fig. 126   Fig. 127   Fig. 128

11. Determine the angle θ (Fig. 124).

12. Determine the radius *r* (Fig. 125).

13. Three holes *A*, *B* and *C* are to be bored in a plate as shown in Fig. 126. Calculate the distance between *B* and *C*, and the diameter of the pitch circle.

14. Calculate the depth of the groove (Fig. 127).

15. Calculate the depth d (Fig. 128).

16. A hole is to be bored along the line XY (Fig. 129). Calculate the distance d.

17. A hole is to be bored through a block along the line XY. Calculate the angle θ (Fig. 130).

18. Determine the distance d (Fig. 131).

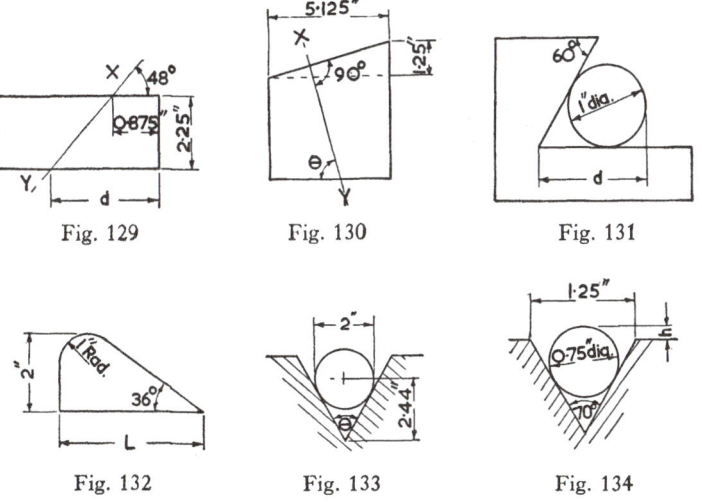

Fig. 129   Fig. 130   Fig. 131

Fig. 132   Fig. 133   Fig. 134

19. Calculate the length L of the profile gauge (Fig. 132).

20. Determine the angle θ (Fig. 133).

21. A cylinder of diameter 0·75 in. rests in a V groove of width 1·25 in. The angle of the groove is 70°. Calculate the dimension h (Fig. 134).

22. A ramp AB to an air-raid shelter is 40 ft. long, and is inclined at an angle of 22° to the horizontal. Determine the depth of the floor below ground level.

To increase the depth to 20 ft. below ground level, how much longer would the ramp have to be made (Fig. 135)?

23. Calculate the dimensions X and Y (Fig. 136).

24. Determine the dimension F in Fig. 137 for the hexagonal nut.

25. Calculate the dimension X (Fig. 138).

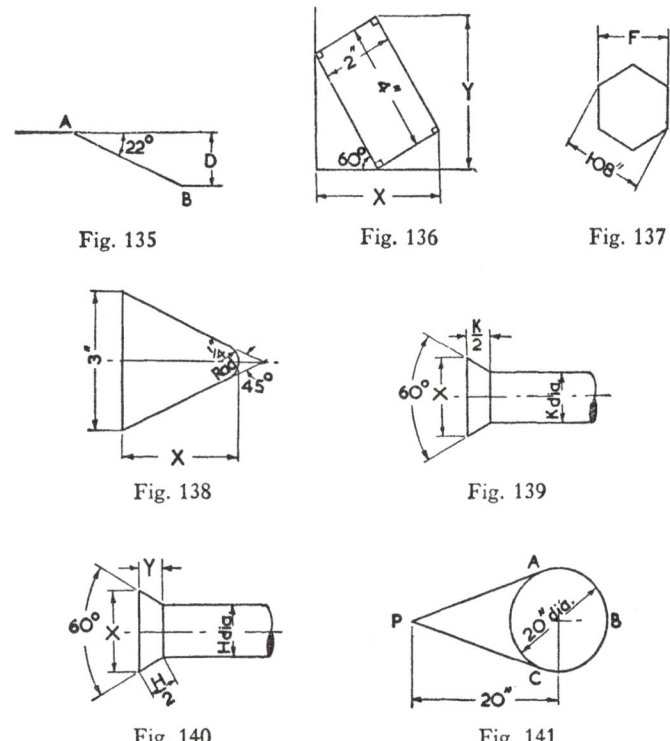

Fig. 135   Fig. 136   Fig. 137

Fig. 138   Fig. 139

Fig. 140   Fig. 141

26. For the Countersunk Flush Head Bolt (Fig. 139) calculate the diameter X for the cases when $K=1$ in., $\frac{3}{4}$ in. and $\frac{1}{2}$ in.

27. Calculate the dimensions X and Y for the bolt shown in Fig. 140 for the cases when $H=\frac{7}{8}$ in., $\frac{5}{8}$ in. and $\frac{3}{8}$ in.

28. An endless belt passes round a small peg P, and the portion ABC of a wheel of diameter 20 in. Calculate the length of the belt (Fig. 141).

Exercise 54. Trigonometry: Miscellaneous Problems

1. A surveyor measures a distance of 740 ft. up a slope of 15° to the horizontal. What length must he plot on a map?

2. A surveyor finds the distance between two points is 55 ft., and the difference in height of the two points is 7 ft. What angle does the ground make with the horizontal?

3. A railway track rises 1 in 180. What is the angle which the track makes with the horizontal?

4. A ladder is placed against a window sill and is inclined at an angle of 19° to the vertical. If the foot is 8 ft. from the wall calculate the height of the window sill.

5. Calculate the horizontal drift of a balloon if its cable is 3000 ft. long and makes an angle of 75° with the horizontal.

6. The ridge of a roof is 8 ft. above the eaves, and the half-span is 17 ft. Calculate the inclination of the roof to the horizontal.

7. On a staircase the tread is 11 in. and the rise is 8 in. Determine the inclination of the stairs to the horizontal.

8. Determine the length of one of the beams of a roof if the roof makes an angle of 35° with the horizontal and the half-span of the roof is 30 ft.

9. A horse is towing a barge along a canal. If the tow rope makes an angle of 25° with the direction of motion of the barge and the horse exerts a force of 700 lb., determine the force pulling the barge along.

10. A hill is ¼ mile long and makes an angle of 32° with the horizontal. Calculate the height of the hill in feet and also the length of the hill as it would be shown on a map.

11. An aeroplane flies N. 17° E. for 20 min. at a speed of 200 m.p.h. How far north of its starting point will it be then?

12. The height of a church spire is found by measuring the angle of elevation from a point 150 ft. away. If the angle is 32° 23′, calculate the height of the spire.

13. A surveyor measures the width of a river by measuring a line AB of 93 ft. along the bank of the river, and then finds the bearing of a point C on the bank opposite to A. If the angle ABC is 74° 51′, calculate the width of the river.

14. If a roof makes an angle of 28° with the horizontal and the beams are 19 ft. long, calculate the height of the ridge above the eaves.

15. If a pole 7 ft. 6 in. long throws a shadow of 5 ft. 10 in., determine the angle of elevation of the sun.

16. A pendulum of length 20 cm. swings on either side of the vertical through an angle of 15°. Through what height does the bob rise?

17. A ladder 20 ft. long is placed against a window sill 17 ft. high. What is the angle of inclination of the ladder to the horizontal?

18. An aeroplane flies on a course N. 29° E. for 15 min. at 180 m.p.h. Find how far east of its starting point it is then.

19. A V groove has an angle of 33° and the width across the top is 1·47 in. Calculate the depth of the groove.

20. A plane is flying directly overhead at a speed of 180 m.p.h. Fifty seconds later the angle of elevation of the plane is 20° 17′. If it is flying horizontally determine its height.

ANSWERS

Exercise 1 (p. 1)

1. (a) 73 ft. $10\frac{1}{4}$ in.
   (b) 37 ft. 0 in.
   (c) 62 ft. 11 in.
   (d) 15 ft. 0 in.

   (e) 108 tons 10 cwt.
   (f) 56 tons 10 cwt.
   (g) 165 tons 0 cwt.
   (h) $\frac{113}{330}$.

2. (a) 61 ft. 2 in.
   (b) 51 ft. 2 in.

   (c) 127 tons 13 cwt.
   (d) 900 gal.

3. 1 ft. 6 in., 1 ft. 0 in.

Exercise 2 (p. 3)

1. $L = 13\frac{13}{16}$ in.
2. $X = 18\frac{5}{16}$ in.
3. $d = 4\frac{1}{8}$ in.
4. $D = 3\frac{15}{32}$ in.
5. $L = 2\frac{1}{16}$ in.
6. $R = 1\frac{3}{16}$ in.
7. $L = 3\frac{3}{64}$ in.
8. $X = 1\frac{19}{64}$ in.
9. $L = 1\frac{63}{64}$ in.
10. $d = 1\frac{13}{32}$ in.
11. $X = 1\frac{21}{32}$ in.
12. $L = \frac{25}{32}$ in.
13. $R = \frac{3}{8}$ in.
14. $X = 1\frac{3}{4}$ in.
15. $L = 6\frac{19}{64}$ in.
16. $X = 1\frac{3}{64}$ in.
17. $a = \frac{27}{32}$ in., $b = 4\frac{1}{16}$ in., $c = \frac{17}{64}$ in.
18. $A = 3\frac{1}{4}$ in., $B = \frac{51}{64}$ in., $C = 5\frac{29}{64}$ in.
19. $h = 2\frac{5}{8}$ in.
20. $d = 1\frac{1}{2}$ in.

Exercise 3 (p. 7)

1. $10\frac{1}{2}$ sq.in.
2. 9 cu.in., $2\frac{1}{4}$ lb.
3. $2\frac{3}{8}$ lb.
4. $1\frac{31}{112}$ in.
5. $\frac{11}{70}$ lb.
6. $\frac{13}{48}$ cu.in.
7. $4\frac{3}{8}$ lb.
8. $\frac{7}{32}$ in.
9. $\frac{7}{32}$ in.
10. $1\frac{1}{4}$ in.
11. $2\frac{1}{5}$ in.
12. 6.
13. $\frac{1}{2}$ in.
14.

d	$\frac{3}{4}$ in.	$\frac{7}{8}$ in.
$2d$	$1\frac{1}{2}$ in.	$1\frac{3}{4}$ in.
$\frac{3}{4}d$	$\frac{9}{16}$ in.	$\frac{21}{32}$ in.

15. $\frac{7}{8}$ in., $1\frac{1}{16}$ in., $2\frac{3}{8}$ in.
16. $6\frac{5}{12}$ lb.

Exercise 4 (p. 9)

1. 672 lb., 1120 lb., 448 lb. 2. 1215 lb.

3. $176\frac{16}{19}$ lb. 4. 5 in. 5. $\frac{3}{4}$. 6. 90 tons.

7. $\frac{103}{173}$. 8. 36 ft. 9. £360, £336.

10. 3 tons, 48. 11. 5600 metres, $\frac{15}{32}$ mile. 12. 20, 1 in.

13. $4\frac{4}{5}$ lb. 14. $1\frac{1}{4}$ in. 15. $\frac{9}{16}$ in.

Exercise 5 (p. 10)

1. (a) 3663·5 sq.ft. (b) 2009·3 sq.ft.
   (c) 1607·0 sq.ft. (d) 1486·25 sq.ft.

2. 252·7 sq.ft. 3. 111·2 sq.ft. 4. $L = 2·125$ in.

5. $L = 16·8125$ in. 6. $D = 8$ in., $d = 4·875$ in.

7. $D = 2·375$ in., $X = 0·548$ in. 8. $L = 1·375$ in.

9. $X = 1·4592$ in. 10. $L = 2·074$ in., $D = 0·565$ in.

11. $a = 3·36$ in., $b = 0·725$ in., $c = 5·485$ in.

12. $R = 1·0785$ in. 13. $X = 6·0126$ in., $D = 0·7374$ in.

Exercise 6 (p. 13)

1. (a) 1·0505. (b) 97·4087. (c) 0·0245. (d) 0·0278.

2. 14 sq.in. 3. 12·2094 cu.in. 4. 0·94 ft.

5. 440. 6. 3·78125 sq.in. 7. 4·25 sq.in.

8. 16·7 in., 13·5 sq.in. 9. 2·27 in. 10. 4·988 sq.in.

11. 19·76 in. 12. 9·14 sq.in. 13. 54·26 in.

14. 12·5. 15. 1·176.

Exercise 7 (p. 15)

1.

d	$1 \cdot 7d$	$0 \cdot 6d$
0·75	$1\frac{1}{4}$	$\frac{7}{16}$
0·875	$1\frac{1}{2}$	$\frac{1}{2}$
1·125	$1\frac{15}{16}$	$1\frac{1}{16}$

2.

d	$1 \cdot 6d$	$1 \cdot 45d$	$0 \cdot 7d$
1	$1\frac{19}{32}$	$1\frac{7}{16}$	$\frac{11}{16}$
0·875	$1\frac{13}{32}$	$1\frac{9}{32}$	$\frac{5}{8}$
1·125	$1\frac{13}{16}$	$1\frac{5}{8}$	$\frac{25}{32}$

3.

d	$2d$	$0 \cdot 75d$
0·75	$1\frac{1}{2}$	$\frac{9}{16}$
0·625	$1\frac{1}{4}$	$\frac{15}{32}$
1·125	$2\frac{1}{4}$	$\frac{27}{32}$

4. (a) $\frac{5}{32}$. (b) $\frac{7}{32}$. (c) $\frac{3}{16}$.

5. (a) $\frac{7}{64}$. (b) $\frac{1}{8}$. (c) $\frac{11}{64}$.

6.

	D	t
(a)	$2\frac{19}{64}$	$\frac{11}{32}$
(b)	$2\frac{1}{64}$	$\frac{19}{64}$
(c)	$3\frac{47}{64}$	$\frac{9}{16}$

7. $1 \cdot 5d = 3\frac{15}{16}$, $\quad 0 \cdot 5d = 1\frac{5}{16}$,
$2 \cdot 625d = 6\frac{57}{64}$, $\quad 1 \cdot 25d = 3\frac{9}{32}$,
$1 \cdot 125d = 2\frac{61}{64}$.

8. $\frac{17}{32}$ in.   **9.** $4\frac{5}{64}$ ft.   **10.** $22\frac{13}{16}$ in.

Exercise 8 (p. 16)

1. 4 ft.   **2.** 44.   **3.** 8 ft. 8 in.   **4.** 157.

5. 5 ft. 3 in.   **6.** 15.   **7.** 64.   **8.** 88.

9. $3\frac{1}{8}$ in.   **10.** $\frac{1}{2}$ in.

Exercise 9 (p. 17)

1. 14.   **2.** -2.   **3.** -3.   **4.** -16.   **5.** 4.

6. -22.   **7.** -1.   **8.** 6.   **9.** 32.   **10.** -8.

11. 6.   **12.** -8.   **13.** 0.   **14.** 0.   **15.** -1.

16. 20.   **17.** 2.   **18.** 11.   **19.** -36.   **20.** 0.

Exercise 10 (p. 18)

1. 4.   **2.** 10.   **3.** −6.   **4.** −13.   **5.** 22.

6. −31.   **7.** −12.   **8.** 19.   **9.** −61.   **10.** −18.

11. 4.   **12.** 16.   **13.** −2.   **14.** 21.   **15.** −20.

16. 6.   **17.** 10.   **18.** 40.   **19.** −19.   **20.** 0.

Exercise 11 (p. 18)

1. 15.   **2.** −21.   **3.** −24.   **4.** 16.   **5.** −14.

6. −28.   **7.** 72.   **8.** −40.   **9.** −12.   **10.** 1.

11. 35.   **12.** −21.   **13.** −54.   **14.** 30.   **15.** 24.

16. −80.   **17.** −90.   **18.** −6.   **19.** 24.   **20.** −1.

Exercise 12 (p. 19)

1. 3.   **2.** −7.   **3.** −3.   **4.** 5.   **5.** −6.

6. −5.   **7.** −4.   **8.** 3.   **9.** −3.   **10.** 1.

11. $\frac{6}{35}$.   **12.** $-\frac{15}{28}$.   **13.** $\frac{40}{63}$.   **14.** $-\frac{15}{28}$.   **15.** $\frac{63}{100}$.

16. $\frac{24}{77}$.   **17.** $\frac{1}{2}$.   **18.** 500.   **19.** −1600.   **20.** 6000.

Exercise 13 (p. 19)

1. $8x$.   **2.** $10a$.   **3.** $2b$.   **4.** $3c$.   **5.** $-20x$.

6. $16x$.   **7.** $8a$.   **8.** $5b$.   **9.** $-11x - 6y$.

10. $4a + b + 8c$.   **11.** $4a - b + 5c$.   **12.** $5a^2 - 8a - 11$.

13. $a^2b + 3ab^2 + 5ab$.   **14.** $9u - 2v - 8w$.   **15.** $6a - b + 4c$.

16. $9x^3 + 4x^2 + x - 6$.   **17.** $6a + 7b$.   **18.** $16a - 6b - c$.

19. $8x + 4b$.   **20.** $4y - 3x$.

75

Exercise 14 (p. 20)

1. $3a$.   2. $5ab$.   3. x^2y^2.   4. $-6ax$.   5. $a-2b$.

6. $-2a-3b+12c$.   7. $18a^2-2a-13$.   8. $ab+12ab^2$.

9. $8u-w$.   10. $-3a-7b+3c$.   11. $-a+10b+9c$.

12. $3a-11b+4c$.   13. $a^3-5a^2b+3ab^2+ab^3-b^3$.

14. $-2a-6b+14c$.   15. $-2a^2-15a+10,\ 16a^2+a+2$.

16. $6u-5k-15v$.   17. $12x-a-b$.   18. $S-xy$.

19. $x^2-3xy-2y^2$.   20. $2x+4y$ south of A.

Exercise 15 (p. 21)

1. $6x^2$.   2. $15x^3$.   3. $12a^5$.   4. $4m^5$.   5. $18a^3b^3$.

6. $6a^5b^6c^4$.   7. $18u^3v^3$.   8. $-24a^3b^3$.   9. $8a^3$.

10. $-24x^4y^4$.   11. a^3-2a^2-3a.   12. $2x^4-3x^2+x$.

13. $4y^3-6y^2+12y$.   14. $c^5-c^4-c^3+c^2$.   15. $9a^4b^2-6a^3b^3$.

Exercise 16 (p. 22)

1. x^2+6x+8.   2. a^2+a-12.   3. $a^2-8a+15$.

4. $a^2+3a-70$.   5. $6a^2-11a-10$.   6. $15x^2-47x+28$.

7. $4c^2-26c+30$.   8. x^2+4x+4.   9. $x^2-8x+16$.

10. $4x^3-12x^2+17x-12$.   11. $6x^2+17xy+12y^2$.

12. $9x^2-4y^2$.   13. $2x^2-4xy-6y^2$.   14. $\pi(R^2-r^2)$.

15. $\dfrac{\pi}{4}(D^2-d^2)$.   16. $15a^2$ sq.ft.   17. $9x^2$ sq.in.

18. $34x^2$.   19. $60x^2$ sq.ft.   20. $19a^2$.   21. $x^3+6x^2+11x+6$ cu.in.

6

Exercise 17 (p. 23)

1. x^3. **2.** $-a^5$. **3.** $-x^4$. **4.** a. **5.** $-5c^2$.

6. $-8x$. **7.** $7a^6$. **8.** $-7d^3$. **9.** 3. **10.** $\frac{1}{2}a^4$.

11. $\frac{1}{x}$. **12.** $-\frac{1}{a^4}$. **13.** $-\frac{1}{b^3}$. **14.** $\frac{1}{c^2}$.

15. x^2y^2. **16.** $-a^4b^2$. **17.** c^2d^3. **18.** $4a^3b^4c^5$.

19. $\frac{5}{xy^2z^3}$. **20.** $\frac{3ac^3}{b^2}$.

Exercise 18 (p. 23)

1. 4. **2.** 3. **3.** 3. **4.** 3. **5.** 4.

6. 1. **7.** 0. **8.** -3. **9.** -4. **10.** 4.

11. $\frac{1}{2}$. **12.** $\frac{2}{3}$. **13.** $\frac{5}{7}$. **14.** $\frac{1}{4}$. **15.** $-\frac{1}{3}$.

16. $-\frac{1}{2}$. **17.** $\frac{1}{5}$. **18.** $\frac{1}{2}$. **19.** -1. **20.** 0.

21. $\frac{1}{4}$. **22.** $\frac{1}{15}$. **23.** $\frac{3}{8}$. **24.** $\frac{1}{5}$. **25.** $-\frac{3}{35}$.

26. $-\frac{1}{16}$. **27.** $\frac{1}{8}$. **28.** $\frac{5}{6}$. **29.** $-\frac{3}{16}$. **30.** $\frac{1}{96}$.

31. $1\frac{1}{2}$. **32.** 8. **33.** $\frac{9}{14}$. **34.** $\frac{3}{4}$. **35.** $1\frac{1}{8}$.

36. $2\frac{2}{9}$. **37.** $-\frac{1}{2}$. **38.** $-\frac{1}{4}$. **39.** $\frac{1}{3}$. **40.** 1.

Exercise 19 (p. 24)

1. 2. **2.** 3. **3.** 2. **4.** 2. **5.** 7.

6. 3. **7.** 3. **8.** $\frac{4}{7}$. **9.** 3. **10.** 4.

11. 5. **12.** 5. **13.** 2. **14.** 2. **15.** 6.

16. 5. **17.** 5. **18.** 2. **19.** 2. **20.** $3\frac{2}{3}$.

21. 13. **22.** 12. **23.** 6. **24.** $3\frac{3}{7}$. **25.** 18.

26. $2\frac{14}{17}$. **27.** $7\frac{1}{2}$. **28.** 20. **29.** 36. **30.** $1\frac{1}{5}$.

31. $\frac{8}{21}$. **32.** $\frac{8}{9}$. **33.** $\frac{3}{4}$. **34.** 1. **35.** $10\frac{1}{2}$.

Exercise 20 (p. 25)

1. -4.   2. -3.   3. $\frac{6}{7}$.   4. $\frac{4}{7}$.   5. 7.

6. 3.   7. -7.   8. -4.   9. 6.   10. 2.

11. 1.   12. $3\frac{1}{9}$.   13. $2\frac{2}{15}$.   14. $\frac{1}{3}$.   15. $-1\frac{5}{14}$.

16. 4.   17. $2\frac{1}{19}$.   18. 5.   19. $\frac{2}{3}$.   20. -27.

Exercise 21 (p. 26)

1. 2.   2. -7.   3. -1.   4. -4.   5. $-1\frac{1}{2}$.

6. 30.   7. 5.   8. 1.   9. $-\frac{7}{36}$.   10. 5.

Exercise 22 (p. 26)

1. $2a$.   2. $3b^2$.   3. 4.   4. 1.   5. $b-a$.

6. $2c$.   7. $a-b+c$.   8. $\dfrac{b}{a}$.   9. $\dfrac{d-a}{c}$.

10. $\dfrac{bc-d}{4}$.   11. $\dfrac{3c-b-d}{a}$.   12. $2bc-d$.

13. $\dfrac{a}{4}$.   14. $2a$.   15. $\dfrac{3b}{2}$.   16. $\dfrac{5cd}{6}$.

17. $\dfrac{5}{3a}$.   18. $\dfrac{2a-2b+3}{9}$.   19. $\dfrac{13}{a}$.   20. $\dfrac{2a+b}{b^2-a}$.

Exercise 23 (p. 27)

1. 8 ft., 11 ft.   2. 11·9 in., 10·1 in.   3. 2 in., 4 in., 5 in.

4. $9\frac{1}{3}$ in.   5. 180 sq.in.   6. 42 ft., 21 ft.

7. 12 cm.   8. 100 ft., 20 ft., 5 ft.   9. 75.

10. $36\frac{2}{3}°$ C.   11. 320° F.   12. 910 ft.

13. 245.   14. 72.   15. $1\frac{2}{3}$ m.p.h.

16. 12×12 in.   17. 65 in., 45 in.   18. 1280.

19. 32.   20. 5.

Exercise 24 (p. 29)

1. 4.	**2.** 4.	**3.** 6.	**4.** 8.	**5.** 12.
6. 81.	**7.** 108.	**8.** 5.	**9.** 256.	**10.** 32.
11. $\frac{1}{16}$.	**12.** $\frac{1}{24}$.	**13.** $\frac{1}{16}$.	**14.** 8.	**15.** $\frac{2}{27}$.
16. $\frac{1}{6}$.	**17.** $1\frac{1}{8}$.	**18.** $\frac{1}{3}$.	**19.** $\frac{1}{4}$.	**20.** $\frac{1}{2}$.
21. 8.	**22.** 16.	**23.** 16.	**24.** $\frac{1}{2}$.	**25.** 108.
26. 12.	**27.** 18.	**28.** 375.	**29.** 32.	**30.** 0.
31. 288.	**32.** 48.	**33.** 0.	**34.** 0.	**35.** $1\frac{3}{4}$.
36. 16.	**37.** -7.	**38.** 38.	**39.** 70.	**40.** 28.
41. 5.	**42.** -96.	**43.** $-\frac{1}{20}$.	**44.** $-\frac{2}{3}$.	**45.** $\frac{2}{9}$.
46. 14.	**47.** 29.	**48.** -118.	**49.** 79.	**50.** 13.

Exercise 25 (p. 30)

1. -6.	**2.** 4.	**3.** -27.	**4.** 9.	**5.** -16.
6. 9.	**7.** 8.	**8.** $+8$.	**9.** $+4$.	**10.** -36.
11. -4.	**12.** -20.	**13.** -3.	**14.** -22.	**15.** 13.
16. -25.	**17.** -6.	**18.** $\frac{1}{6}$.	**19.** 0.	**20.** $-\frac{1}{24}$.

Exercise 26 (p. 30)

1. $L=\dfrac{A}{B}$, 4. **2.** $B=\dfrac{V}{LH}$, 2. **3.** $n=\dfrac{1}{p}$, $\frac{1}{12}$.

4. $L=\dfrac{V}{A}$, $8\frac{1}{3}$. **5.** $K=\dfrac{L-D}{2}$, 0·25. **6.** $C=\dfrac{d+D}{2}$, 2·75.

7. $R=R-R_1-R_2$, 7. **8.** $t=\dfrac{v-u}{g}$, 25. **9.** $r=\sqrt{\dfrac{V}{\pi h}}$, $\frac{1}{2}$.

10. $d=\sqrt{\dfrac{4V}{\pi l}}$, $3\frac{1}{2}$. **11.** $n=2CP-N$, 24. **12.** $n=N-2CP$, 40.

13. $C = \dfrac{SF}{S-P}$, 4.

14. $C = \sqrt{\dfrac{HJ}{RT}}$, 0·5.

15. $T = \dfrac{H}{0·06C^2R}$, $11\frac{1}{9}$.

16. $h = \dfrac{S}{\pi r} - r$, $5\frac{1}{2}$.

17. $r = \dfrac{R}{1+at}$, 200.

18. $R = \dfrac{E}{I}$, $33\frac{1}{3}$.

19. $r = \dfrac{E-IR}{I}$, 20.

20. $d = \sqrt{\left(D^2 - \dfrac{4A}{\pi}\right)}$, $3\frac{1}{4}$.

21. $A = \dfrac{r^2}{0·36}$, 49.

22. $l = \sqrt{(a^2 - c^2)}$, 4.

23. $D = (L-C)^2$, 49.

24. $b = \dfrac{a^2 - c^2}{2a}$, $1\frac{2}{3}$.

25. $g = \dfrac{4\pi^2 l}{T^2}$, 32.

26. $n = \dfrac{2S}{a+l}$, 90.

27. $d = \dfrac{l}{a(n-1)}$, $-\frac{5}{18}$.

28. $P = \dfrac{Ah}{12} + Q$, 4.

29. $r = \sqrt{\dfrac{l}{a}}$, 3.

30. $b = \dfrac{a}{ac-1}$, 8.

Exercise 27 (p. 32)

A

1. $\dfrac{1}{a}$.

2. $\dfrac{2b}{a}$.

3. $\dfrac{3c-2b}{3}$.

4. $\dfrac{5b-3c}{9}$.

5. $\frac{1}{4}(4a+3)$.

6. $\frac{1}{4}(9a-2b)$.

7. $\dfrac{-2a+b+c}{a+2}$.

8. $\dfrac{a(b+1)}{a-c}$.

9. $\dfrac{a(c-2b)}{2(a-b)}$.

10. $\dfrac{2-a}{3-a}$.

B

1. $-\dfrac{7a}{4}$.

2. $\dfrac{2ab-ay}{b}$.

3. $\dfrac{uf}{u-f}$.

4. $\dfrac{M(g-f)}{(g+f)}$.

5. $\dfrac{tc}{a+tb}$.   **6.** $\tfrac{9}{5}C+32$.   **7.** $\sqrt{\dfrac{m}{a+bm}}$.   **8.** $\dfrac{2S-ut}{t}$.

9. $\dfrac{2S-ft^2}{2t}$.   **10.** $\dfrac{x(b-a)}{b+a}$.

Exercise 28 (p. 33)

A

1. 13.   **2.** 15.   **3.** 22.   **4.** 33.   **5.** 63.

6. 94.   **7.** 13·75.   **8.** 53·9.   **9.** 31·62.   **10.** 98·5.

B

1. 1·25.   **2.** 2·96.   **3.** 0·97.   **4.** 0·36.   **5.** 0·29.

6. 0·35.   **7.** 0·25.   **8.** 10·67.   **9.** 7·24.   **10.** 28·3.

Exercise 29 (p. 33)

1. 1·936, 14·06.   **2.** 4·183, 306·2.   **3.** 5·363, 827·0.

4. 8·306, 4758.   **5.** 9·989, 9958.   **6.** 10·71, 13140.

7. 6·915, 2287.   **8.** 9·24, 7288.   **9.** 18·05, 106000.

10. 29·96, 805100.   **11.** 0·9938, 0·9754.   **12.** 0·6249, 0·1525.

13. 0·3972, 0·0249.   **14.** 0·2927, 0·007345.   **15.** 0·251, 0·003969.

16. 0·1767, 0·0009764.   **17.** 0·0866, 0·00005624.

18. 1·001, 1·004.   **19.** 0·1, 0·0001.   **20.** 0·9999, 0·9998.

Exercise 30 (p. 33)

1. (a) 6. (b) 25. (c) 9. (d) 2·88. (e) 5·43. (f) 8·58. (g) 19·8.

2. 5 in.   **3.** $4\tfrac{1}{2}$.   **4.** 1·425.   **5.** $x=3, y=3·928$.

6. $x=8·07, y=7·198$.   **7.** $x=\sqrt{(a^2+b^2)}, y=\sqrt{(a^2+b^2-c^2)}$.

8. $a=3·162, b=3·873, x=2·449$.   **9.** 1·127.   **10.** 25·3.

Exercise 31 (p. 35)

1. 0·866 in. 2. 1·7366 in. 3. 1·101 ft.

4. 3·75 in. 5. 0·146 in. 6. 0·2 in.

7. 0·2843 in. 8. 7·5 in. 9. 6·325 in.

10. 4·583 in. 11. $x = 1·825$ in., $y = 5·399$ in.

12. $d = 12·5$ in., $h = 14·31$ in. 13. 3·125 in.

14. 3·536 in. 15. 4·62 in. 16. 5·958 in.

Exercise 32 (p. 38)

A

1. 0. 2. 1. 3. 2. 4. 3. 5. 6.

6. $\bar{1}$. 7. $\bar{2}$. 8. $\bar{3}$. 9. $\bar{2}$. 10. $\bar{6}$.

11. $\bar{1}$. 12. 0. 13. $\bar{1}$. 14. 4. 15. 0.

B

1. 0·5845. 2. 1·4150. 3. 2·5105. 4. 3·4751. 5. 1·4051.

6. 0·4343. 7. 2·6378. 8. 1·950. 9. 0·1981. 10. 5·7605.

11. $\bar{1}$·7634. 12. $\bar{2}$·8820. 13. $\bar{4}$·699. 14. $\bar{3}$·3838. 15. $\bar{6}$·8573.

C

1. 392·1. 2. 29·57. 3. 0·7474. 4. 1492.

5. 1·154. 6. 0·4056. 7. 2427. 8. 0·009723.

9. 39570. 10. 3·142. 11. 0·0001. 12. 1·024.

13. 2·718. 14. 0·4343. 15. 1000000.

Exercise 33 (p. 39)

1. 761·7. 2. 14340. 3. 9·872. 4. 137·9.

5. 86·22. 6. 351·5. 7. 16490. 8. 18·06.

9. 4996. **10.** 618·1. **11.** 1209. **12.** 1160.

13. 5671×10^6. **14.** 8·648. **15.** 9·093. **16.** 124·8.

17. 8·628. **18.** 1·444. **19.** 5·446. **20.** 565·7.

21. 5·496. **22.** 79·87.

Exercise 34 (p. 39)

A

1. $\bar{2}$·8. **2.** 0·6. **3.** $\bar{1}$·3. **4.** $\bar{3}$·8. **5.** $\bar{2}$·6.

6. 1·5. **7.** 0·6. **8.** $\bar{7}$·46. **9.** $\bar{2}$·82. **10.** 0·65.

B

1. 0·3. **2.** 2·1. **3.** 2·8. **4.** 1·7. **5.** $\bar{4}$·8.

6. $\bar{7}$·8. **7.** $\bar{3}$·7. **8.** $\bar{2}$·77. **9.** 0·77. **10.** 0.

C

1. $\bar{9}$·6. **2.** $\bar{4}$·4. **3.** $\bar{1}$·26. **4.** $\overline{11}$·77. **5.** $\overline{11}$·0428.

6. $\bar{1}$·234. **7.** $\bar{1}$·323. **8.** $\bar{1}$·9883. **9.** $\bar{1}$·4599. **10.** $\bar{1}$·9532.

Exercise 35 (p. 40)

1. 68·6. **2.** 0·8226. **3.** 0·3177. **4.** 59·27.

5. 0·02441. **6.** 0·003364. **7.** 2·617. **8.** 0·5599.

9. 37·85. **10.** 4·212. **11.** 4·625. **12.** 41·16.

13. 0·06046. **14.** 0·06386. **15.** 0·007178. **16.** 0·1956.

17. 0·04846. **18.** 23·02. **19.** 1·0. **20.** 5·0.

Exercise 36 (p. 40)

1. 9·872. **2.** 7·389. **3.** 4373000. **4.** 28·23.

5. 76·32. **6.** 4·588. **7.** 3·868. **8.** 33·52.

9. 34·18. **10.** 44·31. **11.** 0·08231. **12.** 0·00001846.

13. 0·001187.   **14.** 0·4432.   **15.** 0·04043.   **16.** 0·8762.

17. 0·4067.   **18.** 224·9.   **19.** 1·213.   **20.** 0·06695.

21. 3·262.   **22.** 1·2208.

Exercise 37 (p. 41)

1. 110·4.   **2.** 19·27 lb.   **3.** 2·005.   **4.** 1·625 in.

5. 0·04732 amp.   **6.** 0·1478.   **7.** 3·427.   **8.** 8·0729.

9. 5·768.   **10.** 0·1268.   **11.** 34·82.   **12.** 3·155.

Exercise 38 (p. 42)

2. (a) 176.   (b) 44.   (c) 22.   (d) 66.   (e) 55.

3. (a) 45.   (b) 300.   (c) 68·18.   (d) 27·27.   (e) 20·46.

4. 46.55 m.p.h., 68·27 ft. per sec.

5. 29·45 m.p.h.   **6.** 143 miles.

8. (a) 30.   (b) 20.   (c) 67½.   (d) 13·85.

9. 27·17 m.p.h.   **10.** 1 hr. 43 min., 23·3 m.p.h.

11. 93.   **12.** 129, 280.   **13.** 7 ft.

Exercise 39 (p. 43)

1. 223⅛ sq.ft.   **2.** 8 ft.   **3.** 10 chains.   **4.** 3 in.

5. 20¼ lb.   **6.** £12 16s. 8d.   **7.** 64000 lb.

8. 10 ft.   **9.** 450 ft.   **10.** (a) 1⅛ sq.ft., (b) 4$\frac{31}{36}$ sq.ft.

11. 204¼ sq.ft.   **12.** 36·84 in.

Exercise 40 (p. 44)

1. (a) 15⅜ sq.in.   (b) 51·66 lb.   **2.** (a) 42 sq.in.   (b) 141·12 lb.

3. (a) 14$\frac{9}{16}$ sq.in.   (b) 45·435 lb.

4. (a) 39¼ sq.in.   (b) 117·75 lb.

84 ANSWERS

5. (*a*) 18·175 sq.in.   (*b*) 52·344 lb.

6. (*a*) 4$\frac{39}{64}$ sq.in.   (*b*) 15·4875 lb.

7. (*a*) 18$\frac{3}{4}$ sq.in.   (*b*) 63 lb.   8. (*a*) 38 sq.in.   (*b*) 118·56 lb.

9. (*a*) 22$\frac{7}{8}$ sq.in.   (*b*) 76·86 lb.   10. (*a*) 10$\frac{3}{8}$ sq.in.   (*b*) 31·125 lb.

Exercise 41 (p. 46)

1. 662·4 cu.in., 2·389 gal.   2. 3 ft. 8$\frac{4}{9}$ in.

3. 510 sq.ft., 4710 gal.   4. 11 times.

5. 4 ft.   6. 3 ft.   7. 485 lb.

8. 413·6 sq.ft.   9. 8 cu.ft.   10. 16 in.

Exercise 42 (p. 47)

1. 46 sq.in.   2. 1$\frac{1}{2}$ sq.ft., 6 lb.   3. 24·32 sq.in., 0·8445 lb.

4. 3·6 in.   5. 3 ft. 3 in.   6. 19·9 sq.in.

7. 0·6739 lb.   8. 144 cu.in.   9. 41·15 lb.

10. 41·06 in.   11. 6·952 in.   12. 36·41 lb.

Exercise 43 (p. 48)

1. 900 sq.in.   2. 6 sq.ft.   3. 155 cu.ft.

4. 55·4 lb.   5. 0·2956 lb. per cu.in.   6. 6·825 lb.

7. 30·16 in.   8. 0·4 lb. per cu.in.   9. 13 in.

10. 6 in.   11. (*a*) 180 sq.in.   (*b*) 2$\frac{1}{2}$ cu.ft.   (*c*) 2$\frac{5}{6}$ sq.ft.

12. (*a*) 144 cu.ft.   (*b*) 150 sq.ft.

13. 18·25 sq.ft.   14. 126·36 sq.in.

Exercise 44 (p. 50)

1. (*a*) 9·426 in.   (*b*) 14·14 in.   (*c*) 11·79 in.   (*d*) 4·32 ft.

2. 5·967 in.   3. 305·5 times.   4. 44 ft. per sec.

5. 0·6363 ft.   6. (*a*) 70·89 sq.in.   (*b*) 47·19 sq.in.   (*c*) 24·63 sq.ft.

7. 14·93 in.   **8.** 4.   **9.** 9·09 tons.

10. 9·279 lb.   **11.** 3·9275 sq.in.   **12.** 8·64 sq.in.

13. 1 in.   **14.** $\frac{5}{8}$ in.

Exercise 45 (p. 52)

1. 235·65 cu.in., 188·52 sq.in.

2. (*a*) (i) 535·5 sq.ft., 1259 sq.ft.   (ii) 1794·5 sq.ft.
   (*b*) (i) 835·8 sq.ft., 1880 sq.ft.   (ii) 2715·8 sq.ft.
   (*c*) (i) 461·8 sq.ft., 1090·8 sq.ft. (ii) 1551·8 sq.ft.
   (*d*) (i) 306·6 sq.ft., 727·6 sq.ft. (ii) 1034·2 sq.ft.

3. 296 sq.ft., 831 sq.ft., 175 sq.ft., 1302 sq.ft.

4. 12·76 lb.   **5.** 3·989 in.   **6.** 50·53 in.

7. (*a*) 432 in. (*b*) 864 in.   **8.** 2950 in.

9. 0·201 in.   **10.** 219·4 in.   **11.** 121·3 in.

Exercise 46 (p. 53)

2. 373·9 gal.   **3.** 8·3 in.   **4.** 74·96 gal.

5. 10 ft.   **6.** 3426 gal.   **7.** 6·003 ft. per sec.

8. 6·671 min.   **9.** 3181 gal.   **10.** 1558 gal.

Exercise 47 (p. 54)

1. 292·2 lb.   **2.** 20·78 in.   **3.** 2·236 in.

4. 4·9 lb.   **5.** 10 in.   **6.** $366\frac{2}{3}$ ft. per sec.

7. 32·73 ft. per sec., 6·5 in.   **8.** 1 ton.   **9.** 97.

10. 642·85 cu.in.   **11.** 14·3 cu.in.   **12.** $9\frac{7}{64}$ cu.in.

Exercise 48 (p. 56)

1. (*a*) 1232 cu.in., 550 sq.in.   (*b*) 335·9 cu.in., 222 sq.in.
   (*c*) 314·2 cu.in., 204·2 sq.in. (*d*) 125·7 cu.in., 139·1 sq.in.

2. (*a*) 113·1 cu.in., 113·1 sq.in. (*b*) 1150 cu.in., 531 sq.in.
   (*c*) 9204 cu.in., 2124 sq.in.

3. 61·86 lb. 4. 14490. 5. 2·046 ft.

6. 79·56 cu.ft. 7. 8 in.

8. 367·6 cu.ft., 282·8 sq.ft. 9. 1·231 in.

10. $\frac{a}{8}[4b + \pi a]$, ·162·84 cu.ft. 11. 104·5 lb.

12. 3·271 in. 13. 5·154 gal. 14. 2042 lb.

15. 5416. 16. 1524 lb. 17. 55·5 lb. per cu.ft.

18. 0·278 lb. 19. 6·122 in. 20. 2·1 ft.

Exercise 49 (p. 59)

1. 5·23 in., 13·09 sq.in. 2. 4·713 in., 14·14 sq.in.

3. 12·568 in., 18·852 sq.in. 4. 86°, 12 sq.in.

5. 34·04 sq.in. 6. 75°. 7. 6 in.

8. 0·4296 lb. 9. 1·799 in. 10. 6·8 sq.in.

11. (a) 9·426 in. (b) 23·565 sq.in. (c) 12 sq.in. (d) 11·565 sq.in.

12. 0·1134 lb. 13. 4·178 sq.ft. 14. 90°.

15. 104·4 cu.in.

Exercise 50 (p. 62)

1. (i) $a = h \sin \theta$, $b = h \cos \theta$. (ii) $x = b \sin \theta$, $y = b \cos \theta$.

(iii) $b = c \sin \theta$, $a = c \cos \theta$. (iv) $p = r \sin \theta$, $q = r \cos \theta$.

(v) $c = a \cos \theta$, $d = a \sin \theta$. (vi) $x = l \sin \theta$, $y = l \cos \theta$.

(vii) $a = 5 \sin \theta$, $b = 5 \cos \theta$. (viii) $x = 7 \sin 30°$, $y = 7 \cos 30°$.

(ix) $q = 10 \sin 75°$, $p = 10 \cos 75°$.

(x) $c = 9·75 \sin 25°$, $d = 9·75 \cos 25°$.

2.

	$\sin\theta$	$\cos\theta$	$\tan\theta$
(i)	$\dfrac{a}{l}$	$\dfrac{b}{l}$	$\dfrac{a}{b}$
(ii)	$\dfrac{q}{r}$	$\dfrac{p}{r}$	$\dfrac{q}{p}$
(iii)	$\dfrac{y}{h}$	$\dfrac{x}{h}$	$\dfrac{y}{x}$
(iv)	$\dfrac{p}{r}$	$\dfrac{l}{r}$	$\dfrac{p}{l}$

3.

	$\sin\theta$	$\cos\theta$	$\tan\theta$
(i)	$\frac{5}{13}$	$\frac{12}{13}$	$\frac{5}{12}$
(ii)	$\frac{60}{61}$	$\frac{11}{61}$	$\frac{60}{11}$
(iii)	$\frac{24}{25}$	$\frac{7}{25}$	$\frac{24}{7}$
(iv)	$\frac{9}{41}$	$\frac{40}{41}$	$\frac{9}{40}$
(v)	$\frac{3}{5}$	$\frac{4}{5}$	$\frac{3}{4}$

4. $a=5$ in., $b=8\cdot66$ in.   5. $1\cdot732$ in., 1 in.   6. $7\cdot071$ in., $7\cdot071$ in.

7. $p=5\cdot1963$ in.   8. $\cos\theta=0\cdot9426$, $\tan\theta=0\cdot3536$.

9. $\sin\theta=\frac{5}{13}$, $\tan\theta=\frac{5}{12}$.   10. $0\cdot2875$, $0\cdot9579$.

Exercise 51 (p. 63)

1.

θ	$\sin\theta$	$\cos\theta$	$\tan\theta$
15°	0·2588	0·9659	0·2679
20°	0·3420	0·9397	0·3640
30°	0·5000	0·8660	0·5774
45°	0·7071	0·7071	1·0000
50°	0·7660	0·6428	1·1918
70°	0·9397	0·3420	2·7475
75°	0·9659	0·2588	3·7321
33°	0·5446	0·8387	0·6494
47°	0·7314	0·6820	1·0724
62°	0·8829	0·4695	1·8807

2.

θ	$\sin\theta$	$\cos\theta$	$\tan\theta$
0° 18′	0·0052	1·0000	0·0052
5° 27′	0·0950	0·9955	0·0954
83° 38′	0·9939	0·1109	8·9152
47° 45′	0·7402	0·6724	1·1009
89° 24′	0·9999	0·0105	95·49
0° 59′	0·0172	0·9999	0·0172
37° 15′	0·6053	0·7960	0·7604
69° 59′	0·9396	0·3423	2·7445
3° 19′	0·0579	0·9983	0·0580
36° 52′	0·5999	0·8000	0·7499

3. (*a*) 36° 52′.   (*b*) 13° 39′.   (*c*) 48° 35′.   (*d*) 69° 23′.   (*e*) 0° 57′.

4. (*a*) 36° 52′.   (*b*) 23° 22′.   (*c*) 45° 40′.   (*d*) 73° 29′.   (*e*) 89° 53′.

5. (*a*) 45°.   (*b*) 0° 38′.   (*c*) 34° 26′.   (*d*) 69° 40′.   (*e*) 81′ 36′.

6.

	log sin θ	log cos θ	log tan θ
5° 30′	$\bar{2}$·9816	$\bar{1}$·9980	$\bar{2}$·9836
30° 00′	$\bar{1}$·6990	$\bar{1}$·9375	$\bar{1}$·7614
47° 27′	$\bar{1}$·8672	$\bar{1}$·8301	0·0372
62° 19′	$\bar{1}$·9472	$\bar{1}$·6671	0·2801
85° 53′	$\bar{1}$·9988	$\bar{2}$·8567	1·1422
88° 30′	$\bar{1}$·9999	$\bar{2}$·4179	1·5819
89° 12′	0·0000	$\bar{2}$·1450	1·8550

7. (*a*) 15° 42′.   (*b*) 65° 31′.   (*c*) 48° 01′.   (*d*) 14° 14′.   (*e*) 70° 34′.

Exercise 52 (p. 64)

1. $B = 50°$, $a = 3·214$ in., $b = 3·83$ in.

2. $a = 6·877$ in., $b = 9·466$ in., $B = 54°$.

3. $a = 4·676$ in., $b = 7·396$ in., $B = 57° 42′$.

4. $a = 8·8848$ in., $c = 11·96$ in., $B = 42°$.

5. $b = 8·997$ in., $c = 10·98$ in., $B = 55°$.

6. $A = 26° 18′$.   **7.** $A = 51° 04′$.

8. $B = 32° 30′$.   **9.** $c = 7·789$ in., $A = 37° 35′$.

10. $B = 57° 39′$.

Exercise 53 (p. 65)

1. 20° 13′.   **2.** $A = 41° 49′$; $x = 12·944$ in.

3. $x = 1·19$ in., $y = 1·248$ in.   **4.** 41° 06′.

5. $\theta = 74° 29′$, $x = 6·434$ in.   **6.** 6·238 in.

7. 0·9814 in.   **8.** $h = 2·4571$ in.   **9.** 87° 20′.

10. $\theta = 25° 13′$, $x = 3·237$ in.   **11.** 24° 13′.

12. 6·065 in.   **13.** $BC=1·166$ in., $AB=2·758$ in.

14. 1·14 in.   **15.** 0·202 in.   **16.** 2·901 in.

17. 76° 18′.   **18.** 1·366 in.   **19.** 4·0777 in.

20. 48° 24′.   **21.** 0·1359 in.   **22.** 14·98 ft., 13·38 ft.

23. $X=3·732$ in., $Y=4·464$ in.   **24.** 0·9354 in.

25. 3·2177 in.   **26.** 1·577 in., 1·183 in., 0·7887 in.

27.

X	Y
1·3125 in.	0·379 in.
0·9375 in.	0·271 in.
0·5625 in.	0·162 in.

28. 76·54 in.

Exercise 54 (p. 69)

1. 714·8 ft.   **2.** 7° 19′.   **3.** 0° 19′   **4.** 23·23 ft.

5. 776·4 ft.   **6.** 25° 12′.   **7.** 36° 01′.   **8.** 36·62 ft.

9. 634·4 lb.   **10.** 699·5 ft., 1119 ft.   **11.** 63·75 miles.

12. 95·13 ft.   **13.** 343·5 ft.   **14.** 8·92 ft.

15. 52° 07′.   **16.** 0·682 cm.   **17.** 58° 12′.

18. 21·82 miles.   **19.** 2·481 in.   **20.** 4878 ft.

INDEX